basic principles of digital computers

VESTER ROBINSON

Since the first electric computer was constructed by Bell Laboratories in 1940, there have been dynamic and extensive advances in computer technology. At first glance, it may seem difficult for the mind to grasp all the technical details necessary to construct such complicated machinery. Fortunately, however, the basic computer principles remain fairly stable, even though some models are obsolete before they leave the production line. This book is devoted to developing an understanding of these basic principles.

The whole-part-whole concept of education is used in this text. It starts with a broad view of the whole computer field—what the computer is, what it does, and how it functions. It then focuses on the individual parts of a computer—mathematics, circuits, logic, and functional components. Finally, it assembles the parts to form a whole system, analyzes the system in some detail, and develops the principles of programming.

The book is a clearly developed, elaborately illustrated compilation of the basic computer principles presenting computer circuits and representing each circuit in two leading systems of computer logic symbols. Not only are binary, octal, decimal, and hexadecimal number systems analyzed, but conversions among these systems are also illustrated. Functions of computer components are broken down and combined into a sample

(Continued on back flap)

m whose response to coded instruc- is studied.

e reader should have some high l algebra and a basic understanding ectricity and electronics in order to n maximum benefit from studying book. However, the serious student d be able to acquire a considerable ledge of computers from this work without such a background.

study of this material will develop a foundation in the basic principles of l computers. This knowledge is in- ble to persons who plan careers in computer industry. For the person plans a career in the areas of tenance, design, systems analysis, other technical areas, these basic iples are essential for future success.

r Robinson* is the Training Specialist arge of Instructional Systems Devel- nt with the Computer Systems rtment, 3380th Technical School, er Air Force Base, Mississippi. He is ssociate member of IEEE and er of the IEEE Computer Society. also the author of the following published by Reston Publishing any: **Basic Principles of Electricity, Principles of Electronics, Electronic epts, Electrical Concepts,** and **Digital puter Concepts: A Self-Instructional ammed Manual.**

basic principles
of digital computers

to my wife

Florence Belle

*God's miracle gift of love,
kindness, joy, and understanding*

basic principles
of digital computers

VESTER ROBINSON

Reston Publishing Company, Inc., Reston, Virginia 22090
A Prentice-Hall Company

Library of Congress Cataloging in Publication Data

Robinson, Vester.
    Basic principles of digital computers.

    Bibliography: p.
    1. Electronic digital computers. I. Title.
QA76.5.R495   001.6'4'044   73-11316
ISBN 0-87909-066-9

© 1974 by
Reston Publishing Company, Inc.
A Prentice-Hall Company
Box 547
Reston, Virginia 22090

10 9 8 7 6 5 4 3 2 1

Printed in the United States of America.

preface

Scientists have understood the basic principles of computers for more than a century, but the technology which made them practical has only recently been developed. Today we live in a world of computers. Every phase of our lives involves some contact with computer processing.

Computers came into being to meet the need for handling masses of information under increasingly complex conditions. Essentially they have extended the mind of man as much as modern transportation has extended his legs. Modern civilization relies more and more on computers.

Since the first electric computer was constructed by Bell Laboratories in 1940, there have been dynamic and extensive advances in computer technology. When a person with no previous background walks into the presence of an elaborate computer system, he may well wonder how the mind can grasp all the technical details necessary to construct such complicated machinery. Fortunately the basic principles remain fairly stable, even though some models are obsolete before they leave the production line. This book is devoted to developing an understanding of these basic principles.

You should have some background in high school algebra and a basic understanding of electricity and electronics in order to receive maximum benefit from studying this book. However, the serious student should be

v

able to acquire a considerable knowledge of computers from this work even without such a background.

The material is developed in accordance with the *whole–part-whole* concept of presenting a subject. The whole computer field is introduced in a general way by surveying the history of computers and then focusing on how a computer works. From this general concept, the principles of the various bits and pieces are carefully developed. These pieces are assembled into larger and larger components until a complete system has been formed. Then the functions of the complete system are examined in great detail.

A study of this material will develop a firm foundation in the basic principles of digital computers. This knowledge is invaluable to persons who plan careers in any area of the computer industry. In fact, this is valuable knowledge for any career since all industry makes use of computers. Some good computer programmers have never learned these basic principles, but they could be better programmers if they had. For the person who plans a career in the areas of computer maintenance, design, systems analysis, and other technical areas, these basic principles are essential to your future success.

The banker, the business man, the educator, and others who must associate with computers and computer-oriented people will find this material of great help in their work. The book is easy to read, and it will help you to understand the machine as well as enable you to speak the language of the computer people.

No individual ever creates a work such as this without help from many sources. In this case, the author has received information from individuals and computer product manufacturers too numerous to list. A heartfelt thank you to each of them. A special thank you must go to IBM for permission to revise and reproduce portions of their technical material. Other firms that have been generous with their technical data are Digital Equipment Corporation, Bell Laboratories, National Cash Register, Motorola, Telex Computer Products, Texas Instruments, and Tektronix.

Some special people have made direct and valuable contributions to this work and deserve special credit. Matthew I. Fox, president of Reston Publishing Company has encouraged, advised, and assisted in compiling a professional book. The book reviewers—Matthew Mandl, technical writer and industrial consultant, and Thomas Kubala, Dean of Anne Arundel Community College—have made many valuable suggestions for improving the quality and technical accuracy. Philip Marshal of Digital Equipment Corporation made a special effort to supply large quantities of modern technical data.

Vester Robinson

contents

computer development

This *is not* a history lesson in the strict sense of the word. The names and dates are only important in the sense that they are milestones that mark the *progressive steps* toward our present day electronic computer systems. This chapter deals with some background information of the computer industry, and sketches the highlights of the struggle which brought about the birth of the modern computer.

1-1 EARLY AIDS TO CALCULATION

Today, when we refer to a single number, we call it a *digit*. This terminology probably evolved from the fact that the earliest calculating device was the hands, with each digit (finger) representing a number. In the American College Dictionary, we note that the first definition for digit is *finger or toe*. Another interesting fact is the derivation of the word *calculus*. It comes from a Latin word meaning *pebble*. This indicates that some of the more progres-

1

sive early mathematicians gradually evolved from fingers and toes to the use of pebbles to aid their counting efforts.

The Abacus

One of the earliest mechanical devices used in calculations was the *abacus.* This device is composed of several strands of beads stretched across a frame. The general appearance is illustrated in Figure 1-1. Each string of beads re-

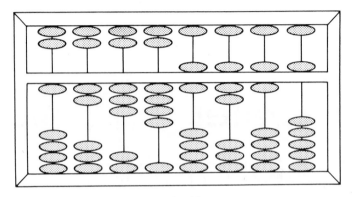

Fig. 1-1. The Abacus.

presents a *place value*: digits, tens, hundreds, thousands, etc.. The number of displaced beads indicate the number of times that place value is to be used. The beads then become columns of figures. The abacus is still popular in some of the eastern countries and skilled operators can manipulate them with amazing speed and accuracy.

The abacus has been *erroneously* called the first computer. *A computer* is a machine which can solve problems and make logical decisions. At the most, the abacus can aid a person in solving his own problems with greater speed and accuracy. It takes no automatic action, and it *does not* make decisions. No, the abacus is not a computer, but it has contributed a great deal to the evolution of numbers and numerical calculators.

Napier's Bones

In 1617, a mathematician by the name of John Napier published a paper in which he described a *mechanical system* of multiplication. He called it *Numbering Rods.* A set of rods consisted of nine boards, and each board had a large number at the top. These numbers were 1, 2, 3, 4, 5, 6, 7, 8, and 9. Each board (rod) contained a vertical column of figures which represented all the multiples of the number at the top. Figure 1-2 shows an example of rods 2, 5, and 8.

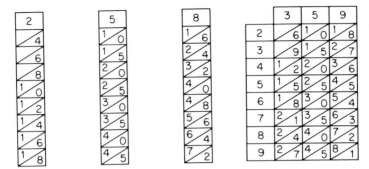

Fig. 1-2. Samples of Napier's Rods.   **Fig. 1-3.** Napier's
Rods with Index.

A *case* contained several sets of these rods along with an *index*. The index consisted of a rack for placement of the rods with a column of figures on the left side. The index figures were 2 through 9. Figure 1-3 shows the index with rods 3, 5, and 9 in position.

The arrangement in Figure 1-3 may be used, if you desire, to multiply the number 359 by any number from 2 through 9. Let's try it. 2 × 359 = 718. The *procedure* is:

1. Select the multiplicand, 2, on the index.

2. Move across to the 9 rod.

3. The number under the 9 is 18. Record the 8 and carry the 1 to the 5 rod.

4. The number under the 5 is now 11. Record 1 to the left of the 8 and carry 1 to the 3 rod. We now have the two least significant digits of the product, 18.

5. The number under the 5 is 6 + 1 which provides the next and final digit of our product. Write the 7 to the left of the 18, and we have 718 which is the product of 2 × 359.

The process may be a bit cumbersome, but in 1617, it was a tremendous aid in the process of multiplication. By following the same procedure, any number, regardless of its size, could be rapidly and accurately multiplied by any digit from 2 through 9. Figure 1-4 shows the arrangement for the number 395,432,698.

Napier's bones were widely used in the 17th, 18th, and 19th centuries. One set that was owned by Charles Babbage is presently *preserved* in the South Kensington Science Museum.

In 1666, another mathematician, Samuel Moreland, invented a machine which replaced Napier's rods with *rotating* multiplier discs. This machine still required some help from the operator in keeping track of carries and addition.

	3	9	5	4	3	2	6	9	8
2	0/6	1/8	1/0	0/8	0/6	0/4	1/2	1/8	1/6
3	0/9	2/7	1/5	1/2	0/9	0/6	1/8	2/7	2/4
4	1/2	3/6	2/0	1/6	1/2	0/8	2/4	3/6	3/2
5	1/5	4/5	2/5	2/0	1/5	1/0	3/0	4/5	4/0
6	1/8	5/4	3/0	2/4	1/8	1/2	3/6	5/4	4/8
7	2/1	6/3	3/5	2/8	2/1	1/4	4/2	6/3	5/6
8	2/4	7/2	4/0	3/2	2/4	1/6	4/8	7/2	6/4
9	2/7	8/1	4/5	3/6	2/7	1/8	5/4	8/1	7/2

Fig. 1-4. Rods Arranged for Multiplication.

Calculators

In 1642, Blaise Pascal invented an *adding* machine which could perform multiplication by a series of additions. In 1694, Wilhelm Leibnitz constructed a machine for *multiplication and division.* Leibnitz's *stepped reckoner* was rather crude and unreliable, but the principles were sound. Some of the parts of the reckoner can still be recognized in modern calculating machines.

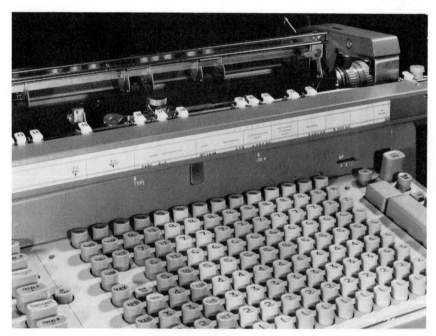

Fig. 1-5. A Modern Adding and Listing Machine. (Courtesy National Cash Register Co.)

During the 18th century *many attempts* were made to build multiplying machines. These efforts met with various degrees of success or failure. The main *hindrance* to such progress was the lack of precision in the mechanical parts of the machines. This was especially true with the cogged wheels and gears.

The first calculator to enjoy *commercial* success was invented in 1820 by Charles X. Thomas. The first *key driven* calculator was patented in 1850 by D. D. Parmalee. In 1872, E. D. Barbour combined a printing device and an adding machine. This produced the first *adding and listing machine.* Our modern cash registers are highly developed adding and listing machines. Figure 1-5 is a picture of a modern adding and listing machine.

1-2 THE APPROACH TO COMPUTERS

The previousy mentioned devices never got beyond the stage of problem solving. Since the machines made *no* decisions, none of them can be called computers. But during the 19th century most of the *principles* of modern computers were fully developed.

Punched Cards

A French inventor, Joseph Marie Jacquard, exhibited an invention in 1801 for *automatic weaving* of fabrics. The Jacquard attachment for looms consisted of a chain of *perforated cards.* As the cards moved through the machine, sinker pegs dropped through the holes to control the weaving process. This enabled the weaving of intricate designs at a tremendous savings of labor. In fact, this system created a mild panic in the weaving industry because the weavers feared they would be replaced by a machine.

Jacquard's system of *punched cards* enjoys such wide use in the modern computer industry that just making standard sized cards is a big business.

Babbage Engines

In 1812, Charles Babbage drew plans for a machine which could solve problems and print answers. The objective of the machine was to calculate and print mathematical tables. He called it the *difference engine,* and his experiments were sponsored by the British Government. The difference engine is on the market today in the form of bookkeeping machines.

In 1833, he started work on another machine which he called the *analytical engine.*

The analytical engine was intended to *store* a thousand numbers of fifty digits each and perform analytical *calculations* in response to stored instructions. Jacquard's punched cards were intended to provide the inputs.

Babbage was too far in advance of the technology of his day. The machine tools were *incapable* of turning out the precision components required to build the machine. The analytical engine was never completed, but the plans and certain working parts are still preserved in science museums.

The Babbage engines were the first *true computers*. They could solve problems and make logical decisions.

Hollerith System

During the census of 1890, an automated system was developed to deal with the mountain of statistics that was gathered. It was a special method of using the Jacquard cards for recording and analyzing information. It was known as the *Hollerith* system, and the Hollerith code is still in use.

The Hollerith *code* divides the 3 × 8 inch card into 12 rows of 80 columns each. Rows 0-9 are punched to indicate corresponding numbers; a hole in row 8 is equivalent to the digit 8. Rows 11 and 12 are normally reserved for control information. Two holes in any given column is a coded letter of the alphabet.

The Hollerith system was used again in the 1900 census. This time, it was so efficient that the results of the census were announced after only one year and seven months. Sound absurd? It wasn't; it was a tremendous step *forward*. Without the system, 100 clerks would have been required to work eight years in order to record only the name, age, and sex of the people listed in that census.

1-3 IMPACT OF ELECTRONICS

As you have probably gathered by this time, the computer was pieced together by technical developments in many different directions. Only in recent years have all these advances been *integrated* into a single computing system.

Electric Power

Even though calculating machines were becoming more popular and more efficient, they were still *just* calculating machines. They were bogged down with *mechanical limitations*. After all, a finite period of time is required for a peg to drop through a hole and open a mechanical switch.

While the builders of these machines were struggling with their mechanical problems, other scientists were incidentally solving their problems for them. In 1750, Benjamin Franklin flew his kite and discovered a new source of power: *electricity*.

Electron Tubes

In 1873, F. Guthrie discharged an electroscope with a piece of hot metal during investigations in *thermionic* emission. Thomas Edison discovered the passage of current from a hot filament to a cold plate, and this led to the invention of *electron tubes*. At the turn of the 20th century Fleming patented his *diode*, and in 1907, Lee Deforest patented the *triode*.

The lid was off and the race was on. Many of the limitations of building computers had been neatly *bypassed*. With electricity for power and electronic switches in the form of tubes, the automatic computer was at last a possibility.

1-4 MODERN COMPUTERS

Special Purpose Computers

In 1940, the first completely *functional* electric computer went into operation. It was developed by Bell Laboratories. The computer is shown in Figure 1-6.

This was intended as a *special purpose* computer; it was designed to repeatedly perform the basic mathematic functions of add, subtract, multiply, and divide. Once in operation, it was discovered to have many other computing capabilities.

The A section of Figure 1-6 shows the relay panel. It contained 450 relays and 10 crossbar switches. Section B shows one of the three remote control stations. This station is a two way teletypewriter for both input and output. The 20 keys contain decimal numbers 0-9 and selected codes including arithmetic signs. The output was printed in hard copy.

This electric relay computer was designed by Bell Laboratory mathematician G. R. Stibitz. It proved to be a successful and important forerunner of our modern computers.

General Purpose Computers

The first completely *functional general purpose* computer was designed by Dr. Howard Aiken and produced through the joint efforts of Harvard University and International Business Machines Corporation in 1944. It was called the Mark I. Figure 1-7 is a picture of this machine.

This machine could solve a *variety* of problems and make logical *decisions*. This meant that the computer had come into its own. The Mark I was rapidly followed by bigger, better, faster, and more reliable computers. In 1946, the ENIAC (Electronic Numerical Integrater and Calculator) solved a

(a)

(b)

Fig. 1-6. The First Functional Computer. (Courtesy Bell Laboratories)

Fig. 1-7. The Mark I Computer. (Courtesy International Business Machines Corp.)

problem in nuclear physics that saved 100 man-years of calculations. In fact, the problem was physically impossible before this time because the problem had many parts which had to be solved in sequence.

Today's Computers

Any reasonably fast computer can solve problems more than a million times *faster* than a highly skilled mathematician. Furthermore, the computer has *complete recall* of the almost limitless information that is stored in its memory banks.

In addition to bringing about the atomic age, computers are now *affecting* the daily lives of every individual. Automation is a fact of the times and automated industry is controlled through computers. While the computer eliminated many jobs through automation, it *created* many new jobs. There are literally thousands of jobs open today that were nonexistent twenty years ago, and new ones are being created all the time. One newspaper estimates that there are *more than* 30,000 desirable jobs open today that have never been listed by the employment agencies.

Today's computers keep our books, operate our factories, guide our space probes, track our satellites, guard our borders from enemy attack, and fly our aircraft. The world of today is so complicated and activity of all types is so rapid, that our way of life would be *impossible* without the computer. The computer has *extended* man's brain as much as radio extended his voice. Computers are everywhere. They are used by everyone from the small business operator to the world government functionary.

Tomorrow's Computers

What will the computer do next? Let your imagination be your guide. It will wash our clothes, cook our food, chart our course to other planets, enforce our laws, drive our automobiles, and perhaps even darn our socks. When we are sick a computer will perform a physical examination, take our temperature, prescribe treatment, and keep a bedside vigilance during our critical period.

The day is near when we will earn and spend our pay check without either check or currency ever passing through our hands. On payday, the bank's computer will deduct a figure equal to salary earned from our employer's account and credit that sum to our account. We will carry a card with our account number. When we make a purchase, the merchant's scanner will transmit our account number, the merchant's account number, and the amount of the purchase to the bank's computer. The computer will subtract that amount from our account and credit it to the merchant's account.

Computers are here to *stay*. They will expand in *two* directions; they will become more *sophisticated*; and they will become *simpler*. The large, sophisticated systems will direct world affairs and activities in space. The small, simple systems will be adopted to *personal* use.

Solid State Computers

Modern computers are being constructed with *solid state components* instead of electron tubes. Transistors and integrated circuits have reduced a mountain of slow, unreliable, high powered machinery to an infinitesimal, highly reliable, low powered, unbelievably fast computer.

1-5 COMPUTER TYPES

All computers fit into one of two categories: analog or digital.

Analog

Most of the history in this chapter relates to the digital computer. The analog computer traces its development along slightly different lines. The slide rule bears the same relation to the analog that our abacus does to the digital. Analog computers deal with *physical quantities*. The quantities are magnitudes and are determined by incremental variations of voltage. When a quantity is expressed with reference to a known standard we call it an *analog*. Analog computers perform addition, subtraction, multiplication and division. They also render logical decisions. They are programmed according to a specific mathematical formula, and they continuously solve a problem with variables in terms of time, quantity, speed, and other calculus-related measurements. These computers are ideally suited to railway switching terminals, gun laying systems, and aircraft guidance.

Digital

While our world is a world of *analog*, we have acquired the habit of expressing our ideas in terms of *digits*. For this reason, a computer which deals with numbers is more flexible. It can be programmed in a digital language to solve *any* problem that we can express in numbers. This is the *digital computer*, and this is the subject of this course. Specifically, we are interested in a completely *automatic* digital computer.

The digital computer also provides greater accuracy at the same cost and produces results at higher speeds than the analog computer.

1-6   AUTOMATIC DIGITAL COMPUTER

This type of computer is not a black box which operates independently of outside influence. Such a device would be of little value. The automatic digital computer is a combination of several functional sections integrated into a compact system. Each functional section has its own specific purpose and each performs its specialized function. The system is designed for each section to complement the other sections in harmonious interrelationships. From this point on, when we use the single word computer, we will be referring to a completely automatic digital computer system.

Stored Program

Our computer will be a computer which *stores* its program internally. *A program* is composed of a group of *coded words* which directs the computer step by step to the solution of a particular problem or the performance of a specific task. We call these coded words *instructions*. In order to operate automatically under program control, the program must be stored internally and be readily available to the *control circuits* which supervise all operations.

Automatic Operation

Once a program has been stored and operation started, the machine is completely *automatic*. It follows the program sequence, decodes the instructions one at a time, and performs as instructed. Most general purpose control programs *do not* terminate. The final instruction on the sequence recalls the first instruction and starts the program over again. The computer *cycles* through the program indefinitely if there is no outside intervention.

1-7 FUNCTIONAL SECTIONS

Nearly all computers are configured into *five* functional sections. These are:

1. Input
2. Memory
3. Arithmetic } = Computer
4. Control
5. Output

These sections are frequently called *units*. The arithmetic section (or unit) is sometimes *erroneously* referred to as the computer because this is the area where calculations are performed. In reality the arithmetic section alone can do nothing. The computer is composed of *all five* sections integrated into a system.

More often than not, the five sections are more *logical* than physical. All five may be contained in a single cabinet and the circuits are usually intermingled for maximum economy of space and parts. The interaction of the five sections is illustrated in Figure 1-8.

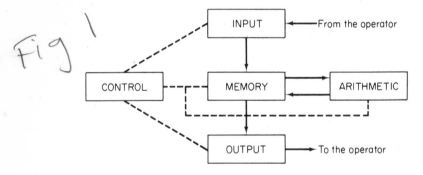

Fig. 1-8. Computer Sections.

Input Section

This is the equipment which enables us to communicate with the machine. All information passing from *man to machine* is processed through this section. There are many types of input devices to facilitate this communication. Card readers, typewriter keyboards, telephone lines, and magnetic tape are some examples of *input devices*. Some computers have the ability to accept voice input. In this case, the microphone is an input device.

The type of input device selected at any given time is determined by the nature of information to be transferred. The input device translates all input information into *digits* and expresses these digits in *electric impulses* of voltage or current. The most widely used computer language is composed of *binary digits*. The binary system has *only two* digits: zero (0) and one (1). A string of these digits readily convert into electric impulses. Each digit is called a *bit*. A *one* bit becomes a *pulse*, and a *zero* bit becomes the *absence of a pulse*. This is illustrated in Figure 1-9. These electronic numbers are passed from the input device to the memory section.

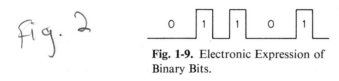

Fig. 1-9. Electronic Expression of Binary Bits.

Memory Section

This is a *storage* area comparable to a rack of boxes in our post office. The program is stored here along with all data essential to the performance of the program. Here we have the *two* types of information which circulate in the computer: *program and data*. Program information consists of *instructions* to the machine and the location of data associated with each instruction. The word data is plural for datum, and it is used to express all types of *facts and figures* except program information. There is a *one way* passage of information from input into memory and from memory to the output section. There is a *two way* exchange of information between memory and arithmetic section.

The incoming electric impulses *magnetize* small particles of ferrite material. The pulses are then *stored* in these minute magnetic fields until they are needed. They can be retrieved by *reversing* the magnetic flux. This type of storage will retain information indefinitely even with the equipment disconnected from the power source. We have a name for a memory which does not lose information when power is removed; we call it a *nonvolatile* memory.

Arithmetic Section

This is the portion of the computer which performs all *calculations* and makes most of the *decisions*. It contains circuits for mathematical operations, shifting, complementing, comparing, counting, etc.. It manipulates the data from memory as directed by the instructions and temporarily holds the results of each operation. When directed to do so, it releases the results back to memory. The nature of the operations in the arithmetic section (calculations, manipulations, and decisions) has led some manufacturers to call this section the computer.

Output Section

This is the *communications link* from the machine to the outside world. Like the input, there are many types of devices for outputting information. Card punch, typewriter keyboard, magnetic tape, magnetic drums, printers, and display cathode ray tubes are a few examples of output devices. The output section accepts pulse type information from the memory section, translates it into understandable form, and presents it to the human operator.

Control Section

Without the control section, the rest of the computer is just so much useless hardware. It provides the timing and control signals which cause all the other sections to function as a computer system. The control section contains the *master oscillator* which is the basis of all timing in the system. It *decodes* the instructions and generates the *control signals* which cause the instructions to be performed exactly as intended. It *initiates* all movement of data and *routes* the information to the proper place.

Typical System

A system may be as simple as a single table top unit or complex enough to require several thousand square feet of floor space. The *complexity* is largely determined by the quantity of storage and the number and types of input output devices. Figure 1-10 illustrates a typical *general* purpose installation.

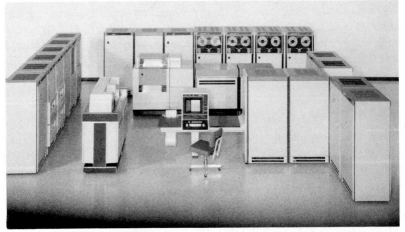

Fig. 1-10. A Modern Computer. (Courtesy National Cash Register)

Hardware Element

The hardware element is the machine. It consists of electronic components which have been selected and arranged according to a predetermined pattern. The pattern, of course, was determined from a list of specifications of what the machine must be capable of doing.

Large scale computers tend to be a bit awe-inspiring, and to some even frightening. The whirring tapes, the clatter of the printer, the unattended typewriter typing out a message, the flashing lights, and the electronic sounds are all part of this mysterious environment.

The computer is actually a *simple* machine composed of a great many repetitions of a few very simple circuits. These simple circuits are interwired in a manner which enables them to react in a prescribed fashion.

Software Element

The *software* element is composed of all the written material that has been composed for or about the hardware. This includes operating manuals, maintenance manuals, programming manuals, and written programs. The software package sometimes costs 25 percent of the hardware price, but without it, the hardware is useless.

The *program* portion of the software usually consists of at least one control program and several subroutines. The *control* program is sometimes called an *executive* program or the *object* program. It may be rather lengthy, consisting of several thousand instructional steps. The control program cycles the computer through a normal routine which causes it to perform all the tasks involved in its *principal* function. Other functions are performed by branching from the control program to any number of *suboutines.* For instance, a subroutine may be used to print out a message, bring in new data, or calculate a mathematical table. When the subroutine task is finished, the program branches back to the control program and the machine continues its normal routine.

Human Element

The *human* element is composed of the men and women behind the machine: designers, builders, installers, programmers, operators, servicers, and technicians, Without the human element, the machine would have never been built. Once in operation, it can function without human intervention, but it is under the control of a human–generated program. Even in automatic operation, the machine accomplishes no real purpose without a human operator.

The operator is *essential,* not for constant control like the driver of an automobile but for maintaining the communications link between man and machine. The operator accepts the outputs and takes appropriate action. He also provides inputs to the machine when such inputs are required.

The operator may be any one of *several* people. The *programmer* fills this roll when he is entering, revising, or analyzing a program. The *maintenance man* serves as an operator when he is interpreting error printouts and diagnosing malfunctions and, of course, there are professional operators whose principal job consists of communicating with the computer.

The operator must be well trained on the procedures for entering and removing information from the machine. He must have some knowledge of the *abilities and limitations* of his computer.

The *programmer* must know how the machine reacts to each instruction as well as be familiar with the capabilities and limitations of the machine. He must be adept at translating problems into meaningful computer language and arranging instructions to solve the problems with the greatest economy of time and storage space.

The *technician* must be adept in the performance of service routines and trouble analysis. It is his job to keep the computer performing at peak efficiency with a minimum of nonoperating (down) time. He needs a more complete knowledge of the detailed functioning of the machine. He must know enough of the operator's job to be able to enter his diagnostic programs and interpret printouts. He must know enough programming to be able to read a program listing and determine the detailed response of the machine to each instruction.

1-8 SUMMARY

1. A *computer* is a machine which performs calculations and makes decisions.

2. The *Babbage engines* would have been true computers, but they were never completed.

3. The word *computer* means a system composed of five functional sections: input, memory, arithmetic, output, and control.

4. A *stored program computer* has its program stored internally for easy, fast access and automatic operation.

5. Two types of information are contained in the computer: *program and data.*

6. An *instruction* is a combination of digits coded to cause a particular operation.

7. A *program* is a list of instructions arranged in the proper sequence for accomplishing a given task.

8. All information is in the form of *digits*, and these digits are translated into *electric impulses.*

9. Data and program information are *separated* in two ways: (a) Lines which carry only one type (b) Common lines with controlled timing sequence (program time and data time).

10. The functional system has three *vital* elements: hardware, software, and human.

1-9 APPLICATIONS OF COMPUTERS

We have mentioned many areas where computers are used and speculated on some of their future applications. Let's move in a little closer and examine some of these applications. The big users of computers have been military, commerce, industry, and science.

Military Applications

The Armed Forces have made *extensive* use of every practical computer system that has been produced. In fact, the military need for better systems (backed by government funds for research and development) has had, and will continue to have, great *influence* on computer development. The Armed Forces of the U.S. is such a huge organization that the smallest task can result in a virtual avalanche of *statistics*. Every type of computer has a military application in *reducing* these statistics to comprehensible terms.

Computers perform such *routine* tasks as maintaining immunization records for every individual. Periodic printouts notify each person of the type of shot he needs and when to report for it. The same procedure is used for keeping track of the individual's accrued leave. Periodic printouts tell each person the number of leave days he has to his credit and how many days he must use to avoid loss of leave time at the end of the fiscal year.

Computers are evident in *every aspect* of the military from the mundane tasks just described to the guidance of space probes. Computers procure and stock supplies, distribute materials to combat forces, keep track of the location of all military forces, ships, aircraft, and weapons. Every aircraft in the air over or near U.S. territory is *tracked* by a computer and identified.

A considerable portion of military personnel have *computer oriented* careers. Jobs range from apprentice repairmen to maintenance personnel to technicians to programmers to systems analysts. In addition to the military personnel, the armed forces provide many *civilian jobs* associated with the computer. These jobs are both civil service and contract maintenance with computer firms.

Commercial Applications

In the field of *commerce*, every large firm makes use of computers, and the small businessman is becoming more and more involved. Computers direct the New York stock exchange and guide the activities of the World Market. *Transportation* companies use computers to direct trucks, trains, ships and aircraft.

Banks would be hopelessly swamped with clerical details without the help of computers. The oddly shaped numbers on your personal check are your account number printed in *magnetic ink*. The computer changes these numbers to voltage waveforms and processes checks several thousand times faster than a clerk would be able to.

This is only one example of thousands of banking tasks which computers perform. Nearly all banks have computer service. Some own computer systems; others lease; and still others acquire computer time by renting.

Industrial Applications

Manufacturing plants use computers for a variety of jobs from payroll computations to complete automation. They control *production* processes, schedules, tools, and assembly lines. They are indispensable in the *design and building* of houses, plants, roads, dams, bridges, and canals. Computers are used to design ships, automobiles, aircraft, space capsules, atomic reactors, and even other computers.

Scientific Applications

All *governments*, *businesses*, and *institutions* involved in scientific ventures use computers. They predict the movement of ocean currents and air masses, weather conditions, and planetary positions. They are used in medicine, astronomy, physics, mathematics, chemistry, biology, biometrics, nuclear physics, and all other areas of research.

Educational Applications

Computer complexes in Denver, Saint Louis and other central locations have computers with *libraries* stored in the memory data banks. Any school in the U.S. that has access to telephone lines and sufficient funds may obtain *computer assisted instruction*. Individual table top input–output devices are available for purchase, lease, or rent.

These units consist of a typewriter keyboard and a cathode ray tube display screen. Students may be entered in a variety of courses. The input–output device is similiar to that shown in Figure 1-11.

This particular unit can be connected to the central computer by telephone lines and provide *personalized* computer instruction for a student. The *display tube* can present text material, schematics, illustrations, etc., from the com-

Fig. 1-11. Graphic Terminal With Copy Unit. (Courtesy Tektronix, Inc.)

puter. The *keyboard* provides the student a means of communicating with the computer to request information, answer questions, etc. The *copy unit* can produce hard copies of any information displayed on the screen. The student could obtain study materials and homework assignments in this way. When the course is completed, the copier can produce a list of *grades and a diploma.*

This type of computer assisted education is due for a phenomenal *expansion.* With the present state of the art, thousands of students can use the same computer on a time sharing basis. Students can be taking different courses or be at different points in the same course. In either case, the access time is so short that each student has the impression that the computer is his own exclusive and *private tutor.*

Personal Applications

Thus far in the history of computers, the *individual* has been bypassed. This is primarily because of two factors: not enough computers and prohibitive cost. This situation is changing. In the near future, computers will be as *common* as television sets. Nearly every home will have computers, and nearly all individuals will be able to have access to a huge computer complex on a purely personal basis.

At least one person has already made computer time *available* to the man on the street. The computer is installed in the lobby of a hotel and a quarter in a slot buys a specified amount of computer time for solving the individual's problem.

The *kitchen* of the future will be completely automated and under computer control. The housewife will input such information as the number of people expected for dinner and the time that dinner will be served. When she returns from her bridge party, golf game, or shopping trip, she will find dinner prepared exactly right, and it will be served up promptly at the designated time.

Review Exercises

1. What is the meaning of the term *digit*?
2. What does the word *calculus* tell us about early computing devices?
3. State two reasons why it is incorrect to refer to the *abacus* as a computer.
4. Define the word *computer*.
5. What was the primary purpose of *Napier's bones*?
6. Describe the main hindrance to the progress of calculating machines during the 18th century.
7. What type of machine first used the Jacquard punched card?
8. Charles Babbage designed the first true computer. *Name* his two machines and *state* the purpose of each.
9. What *two* great ideas are combined in our modern punched card?
10. How does the analog computer *differ* from the digital computer?
11. Describe a computer *instruction*.
12. In what way has the computer *compensated* for the jobs it has eliminated?
13. What is a *program*?
14. A general purpose, stored program computer contains *five* functional sections. List them and state the primary function of each section.
15. What are the *two types* of computer information?
16. How is it possible to *enter* numerical digits into a computer?
17. How is a *one* represented?
18. How is a *zero* represented?
19. What does an *instruction word* contain in addition to specifiying the operation?
20. What is the meaning of the term *data*?

21. Describe two methods of differentiating between program and data information.

22. What are the two general types of computers?

23. List the four biggest users of computers.

24. What two factors have limited individual access to computers?

25. List two advantages of computer assisted instruction over conventional instruction.

computer
operation

Since we are dealing with automatic machines, this chapter is related to a *sequence of functions* rather than being a check list of manual steps. Also we are discussing general purpose computers. This makes our subject so broad that we must assume some specific conditions in order to narrow it down to meaningful limits. The main purpose of this chapter is to form a visual picture of how the computer actions are controlled by instruction words.

2-1 DATA PROCESSING

Data processing is a general term which applies to all manners of moving, storing, analyzing, modifying, and using data information. Some systems are designated as data processors, and these systems may or may not contain a computer. Regardless of the simplicity or complexity of the data processing problem, regardless of who performs the processing (man or machine), certain procedures must be followed.

Data Processing Problem

Let's consider a simple problem such as might be solved once each week by a payroll clerk. The clerk performs a series of operations on the data which pertains to each employee of the firm. The process is something like this:

1. Record vital data on:
   a. Itemized deductions.
   b. Normal hourly wage.
   c. Normal hours worked.
   d. Overtime rates.
   e. Number of hours overtime.

2. Perform these calculations:
   a. Add normal hours worked.
   b. Multiply normal hours by normal hourly wage.
   c. Add overtime hours worked.
   d. Multiply overtime hours by overtime rate.
   e. Add the products obtained in steps b and d.
   f. Add itemized deductions.
   g. Subtract f from e.

3. Write a check for the remainder after step 2g.

The payroll clerk probably does not realize it, but he has been *programmed*. He is a data processor performing a specific job according to a set procedure. As an efficient data processor, he follows the fixed rules and produces fast, reliable, usable results.

The computer is also a data processor. It can be programmed in much the same fashion that the payroll clerk was programmed. If the clerk were replaced by a computer, the only real difference in the results would be the speed of performance. A job that takes the clerk a full 40 hour week will be performed by the computer in a few seconds. The only part of the job that requires any appreciable time is the printing of the checks, and this can be done at the rate of several hundred per minute.

After the programmer has analyzed the problem and recorded the primary machine actions, his listed procedure will look like this:

1. Input and store necessary data on:
   a. Deductions.
   b. Normal hourly wage.
   c. Normal hours worked.

    d. Overtime rates.

    e. Hours overtime.

2. Perform these arithmetic operations on the stored data:

    a. Add normal hours worked.

    b. Multiply a by hourly wage.

    c. Store results of b.

    d. Add overtime hours worked.

    e. Multiply d by overtime rate.

    f. Add the results of b and e.

    g. Store the results of f.

    h. Add deductions.

    i. Subtract h from f.

3. Print out the result of 2i.

Take a close look at the *procedural steps* and determine the *specifications* for a machine to perform this job. It will need an input device; a perforated paper tape would serve our purpose nicely. The machine will need storage space in a random access memory for storage of both program and data; most any type of memory will serve. The machine must be able to add, subtract and multiply; most arithmetic sections can do these things. Our computer needs an output device which can print checks; this machine is a very common piece of hardware.

Nearly any *general purpose* computer is capable of meeting all these specifications and many others. When designing a computer, one of the first steps is to list the specifications that the machine must meet. These specifications are, of course, determined by the types of jobs the computer will be expected to perform.

2-2 SYSTEM BLOCK DIAGRAM

Let's review the computer system block diagram that we constructed earlier. A slight variation of that diagram appears in Figure 2-1.

Input and Storage

The operator enters both program and data information into an appropriate input device. The input device *translates* the facts and figures into electric impulses, and the control section causes them to transfer into designated memory locations. As a general rule, a small area of memory is designated as a *program area* and the program instructions are stored there in sequential order. Each data word has its own memory location, and the words may or

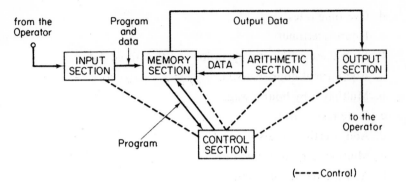

Fig. 2-1. Computer System Block Diagram.

may not be in an orderly sequence. Location of data in memory is of no importance because each instruction contains the address of its particular data and access to all memory locations is *random*.

Starting

The operator *manually inserts* the address of the first instruction of the program into a control panel and presses the start button. The control section accesses the prescribed memory location, removes that instruction word, decodes it, brings the data from memory to the arithmetic section, and generates the required timing and control signals to cause the instruction to be executed.

Automatic Operation

Sequential memory locations in the program area are then accessed to remove, decode, and execute the instructions in the proper order. Provisions are generally made to enable *repetition* of a single instruction or several instructions when the need arises. It is frequently desirable to execute instructions out of the normal sequence. The control unit can be made to take *any instruction* or even jump to a different program when the conditions make such a departure desirable.

Directing the Computer

The control section causes exactly the operations specified by the instructions: no more and no less. The computer does not think; it does not understand; and it has no emotions. The program guides every detail of its operation. The *program* must specify every detail of what to do, what to do it with, when to do it, and how to go about it.

The computer is a high speed machine, nothing more. With proper instructions it can manipulate data, solve problems, and make decisions at a blinding rate of speed. If the programmer provides improper instructions or incorrect data, the computer will follow them, exactly as prescribed, at the same blinding speed. In this case, this high speed robot will manipulate the wrong data, solve the problems incorrectly, make hairbrained decisions, and spit out incorrect information in unbelievable quantities.

2-3 CENTRAL COMPUTER

The portion of the machine that remains when the input and output sections are removed we call the *central computer*. This is the memory section, the arithmetic section, and the control section as illustrated in Figure 2-2.

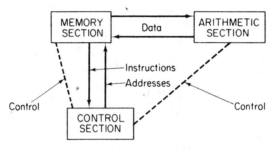

Fig. 2-2. Central Computer.

The majority of the *action* takes place in the central computer: automatically and under program control. Until further notice, we will confine our attention to this central computer area.

Inside the Memory Section

The *storage area* of the memory consists of a group of complete word storage locations. We refer to each location as a *register*. Each register has a specific address and contains a storage device for *each bit* of a complete computer word. The word length is determined during the design phase. Some computers have word lengths as small as six binary (1s and 0s) digits; others have word lengths all the way to 80 digits.

You may *compare* the storage area of memory to a drawer in a filing cabinet. Inside the drawer we find many folders that are numbered sequentially: 1, 2, 3, 4, etc. . Each folder represents a *register* in memory which can contain a complete word. The sequential numbers on the folders are comparable to the memory *addresses* which specify the locations of the registers in memory. Inside each folder, we find several sheets of paper. Each sheet of

Registers

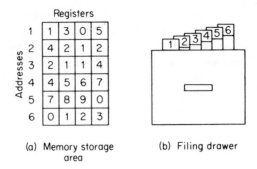

Addresses				
1	1	3	0	5
2	4	2	1	2
3	2	1	1	4
4	4	5	6	7
5	7	8	9	0
6	0	1	2	3

(a) Memory storage area   (b) Filing drawer

Fig. 2-3. Memory Storage Area Compared with a Filing Drawer.

paper corresponds to a *digit* location in the computer word. Each digit has a separate storage device. This analogy is illustrated in Figure 2-3.

For the remainder of this chapter we will revert to the use of common *decimal* numbers to avoid confusion through reference to possibly unfamiliar numbering systems. The first location in this memory is identified by an address of 1. The register at this address holds four decimal digits. The register contents are 1305. This is frequently indicated in abbreviated form in this manner: C(1) = 1305. This abbreviation says, "The contents of location address 1 are 1305."

The second register is located at address 2, and it contains 4212. The third register is located at address 3 and contains 2114, and so on throughout the memory storage area, which may consist of a few hundred to several hundred thousand individually addressable registers.

To expand our filing system, we add more folders. When the drawer is filled, we add other drawers. To increase the memory capacity we add more registers. If one memory unit is not sufficient, we add other memory units until we reach the desired capacity. Figure 2-4 illustrates a memory storage area with 5217 addressable register locations.

This illustration shows the registers stacked in a vertical position. Most central computer memory registers are arranged into a *cube* with rows of storage devices top to bottom, left to right, and front to rear. Sequence of address and layout of the digits in a word is the designer's choice.

Let's suppose that the numbers shown in the designated program area are actually coded instructions. During the computer design phase, any group of numbers can be *assigned* any meaning the designer desires. Let's assign meanings to our numbers. The first two digits are coded instructions and the last two digits are data addresses. Our code is:

10 = clear and add

11 = add

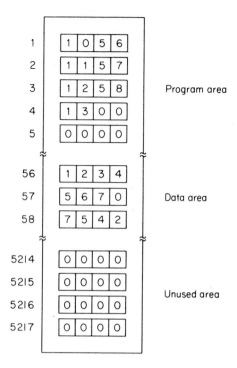

Fig. 2-4. Expanded Storage Area.

12 = store

13 = halt

Now, locations 1, 2, 3, and 4 of Figure 2-4 contain the instructions clear and add, add, store, and halt in that order.

Each instruction must contain an address of the location of the data to be used with that instruction. Examine the last two digits of locations 1, 2, 3, and 4. The clear and add instruction will use the data contained in address 56. The add instruction will use the data from location 57. The store instruction will cause data to be *stored* into location 58. The halt instruction needs no data; therefore, the address portion of this instruction word is normally filled with zeros.

So, our instructions are divided into *two parts*. The first part tells the computer what to do, and the last part gives the location of the data to be used with that instruction. We generally call the first part the *operate section* and the second part the *address section*. In order to have direct access to all registers in our 5217 location memory, we would need to expand the address portion of our instruction word to four digits. That would make matters a bit too complicated for the moment.

Back to Figure 2-4, examine the contents of locations 56, 57, and 58. C(56) = 1234; C(57) = 5670; and C(58) = 7542.

This is an appropriate place to establish the fact that a storage device will always contain a number. Take location 5214 for instance, with a specific number of digits. What does it contain? If you said, "all zeros," you are correct. If you said, "nothing," you are wrong. A zero is a number; it is something. A zero represents a *definite quantity* just as 9 represents a definite quantity. The only difference is in the magnitude; therefore, a zero must be thought of as a number.

Inside the Arithmetic Section

The arithmetic section has many auxiliary circuits such as counters, comparators, and alarms. Right now we will consider the area where calculations are performed and data is manipulated. This area consists of two registers and an adder as illustrated in Figure 2-5.

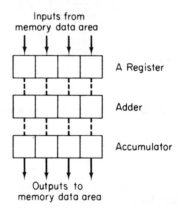

Fig. 2-5. Adder with Associated Registers.

The names of the registers are not important and are likely to have different names with different computers. One register (in our drawing it is the A register) *accepts* data words as they are transferred from memory. This word can now be *added* to the number contained in the accumulator register. The resulting sum will replace the old contents of the accumulator. In order to move a number from a memory storage register into the accumulator register the following steps are used:

1. Move the number into the A register.

2. Clear the contents of the accumulator so that it contains all zeros.

3. Add C(A Reg.) to C(Accum.)

Once a number is in the accumulator, it can be *shifted* left or right, *stored* back into memory, *modified* in several ways, or *moved* into other registers. In some computers, the equivalent of our A register also has shifting capabilities. These registers are sometimes called *dynamic* registers because of their flexible use.

Moving information out of a register *does not* disturb the register contents. This way we can eat our cake and keep it, too. A copy of the information transfers out, but a copy is retained in the register. The information remains in the register until one of three actions destroys it:

1. The register is cleared to zeros.
2. New information is forced in to replace the old.
3. Power is removed from the equipment.

Inside the Control Section

This section contains the *timing* circuits and instruction *decorder*. It *initiates* all movement of information and all actions called for by the instructions. This is done by generating *timing and control* signals and applying them to the proper place at the proper time.

We call the timing oscillators *clocks*. The master oscillator which is the source of all timing is the *master clock*. A large system generally has several auxillary oscillators which are synchronized to the master clock frequency. We call them *slave clocks*. The frequency of the clock system is extremely critical. Both crystal controlled oscillators and tuning fork oscillators have been used successfully.

After the oscillations from the clock have been converted into sharp trigger pulses, they are *distributed* through a counter. The counter has an output from each section with specific time intervals between outputs. The counter can be designed to count off as many pulses as necessary. Each pulse from the counter can then be labeled according to the relative time that it occurs. The time for a complete set of pulses to emerge from the counter is a *cycle*. All actions in the machine are controlled in relation to this timing cycle, and most actions are initiated with specific timing pulses. The timing counter is illustrated in Figure 2-6.

The output of the master clock is a *continuous* string of pulses with sharp edges. The activating portion of these clock pulses is only a fraction of a μs in duration; they are of a uniform amplitude; and the intervals between leading edges are very precise. We have assigned numbers to twelve consecutive pulses to illustrate the generation of a complete cycle of timing pulses.

Let's assign the clock a frequency of 2 MHz. This will give us timing intervals of 0.5 μs and each cycle will be 6 μs in duration. This gives us a

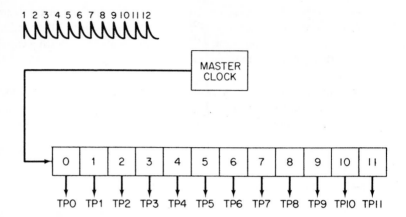

Fig. 2-6. Master Clock and Timing Counter.

relatively *slow* computer but it provides us with concrete facts to use in our discussion.

The procedure is like this:

1. Clock pulse (CP) 1 enters the counter and triggers an output pulse from section 0. We will call this pulse timing pulse (TP) 0.

2. 0.5 μs after CP1, CP2 enters the counter and triggers an output from section 1. This is TP1.

3. 1 μs after CP1, CP3 enters the counter and triggers an output from section 2. This is TP2.

4. The action is repetitive all the way across until CP12 produces TP11. The next clock pulse resets the counter and starts another cycle with TP0.

We are concerned with *four types* of cycles:

1. Memory cycle.
2. Program cycle.
3. Operate cycle A.
4. Operate cycle B.

Each of these are directly related to our master timing cycle, and each has the same 6 μs duration, although they do not start and stop with TP0.

The *memory cycle* is a 6 μs period during which a word is removed from memory or a new word is stored in memory. It is normally triggered with TP1 and automatically terminates 6 μs later. There are *two* distinct types of memory cycles:

1. Memory *storage* cycle.
2. Memory *retrieve* cycle.

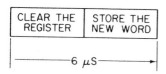

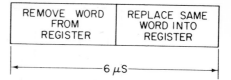

Fig. 2-7. Memory Storage Cycle.

Fig. 2-8. Memory Retrieve Cycle.

Figure 2-7 illustrates the action during the storage cycle.

The address of the register involved was already in place when the memory cycle was triggered. During the first 3 μs, the register is selected, and all bits are *cleared out*, leaving the register with all zeros. Meanwhile, the new word, which comes either from an input device or the arithmetic section, has been moved into position. The final 3 μs of the cycle *moves* this word into the cleared register.

Figure 2-8 illustrates the action during a memory retrieve cycle.

Most memory devices have a *destructive* read out which leaves a cleared register. To avoid loss of information, the word must be *replaced* into the register before it leaves the immediate vicinity.

The *first 3 μs* of the retrieve cycle removes the word from the register and places it into a temporary storage location. This temporary location is normally called the memory information register. During the *final 3 μs* of the cycle, the word is *replaced* in the memory storage register, but a copy is retained in the memory information register. After the replacing action, the word is gated out of the memory information register to either the arithmetic section or to an output device.

The *program cycle* is a 6 μs internal during which an instruction is removed from memory and decoded. This cycle is started by TP0 and ends with TP11. TP0 selects the instruction address, and TP1 starts a memory retrieve cycle to bring out the instruction. The first half of the memory cycle moves the instruction word into the memory information register. At TP7, the word is *restored* to the memory location. TP7 also moves the instruction portion of the word into the *operation* register and the address portion of the word into the *address* register. This matches the content of the memory word which is composed of an instruction and the address of the data to be used with it. This is illustrated in Figure 2-9.

After the instruction portion of the word is *decoded*, the address portion is sent back to memory to obtain the data to be used. The decoding is done by an instruction *matrix* which samples all the digits in the instruction. The matrix then establishes gating levels which combine with timing pulses to trigger the necessary actions at the proper time. We will refer to a basic signal as a *command*. The individual triggers which occur during that instruction we will call *subcommands* or timing pulses.

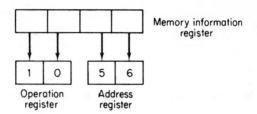

Fig. 2-9. Division of the Instruction Word.

Operate cycle A is the 6 μs interval of time during which data is removed from memory and either used in the arithmetic section or transferred to an output device.

Operate cycle B is the 6 μs interval of time during which a new word is transferred to memory. It can come from either an input device or the arithmetic section.

In our sample machine, we will require *at least* two cycles for most instruction. These will be a program cycle and either an operate cycle A or an operate cycle B. Some instructions will require *all three* cycles. A few instructions will require more time than 18 μs. In this case, we will interrupt a cycle and switch to another form of counting. The interspersing of the cycles for each instruction is shown in Figures 2-10 through 2-13.

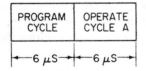

Fig. 2-10. Program and Operate A.

Fig. 2-11. Program and Operate B.

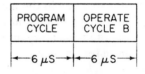

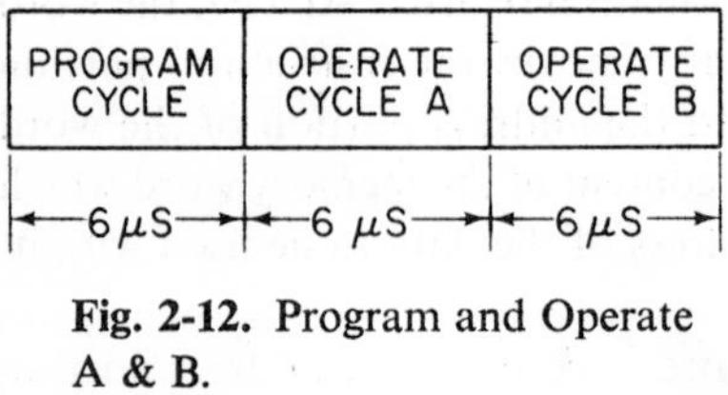

Fig. 2-12. Program and Operate A & B.

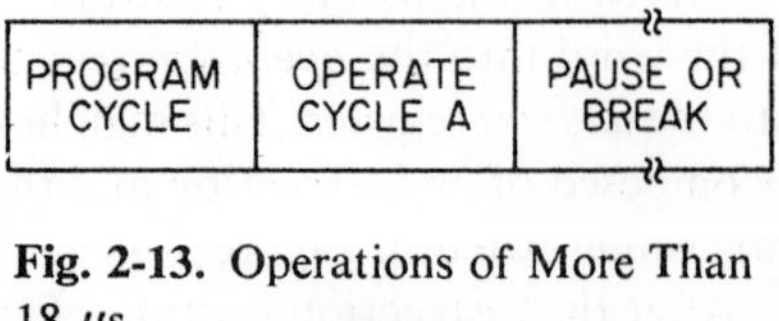

Fig. 2-13. Operations of More Than 18 μs.

A *pause* is used to allow time for extensive arithmetic operations such as multiply and divide. Timing during a pause is accomplished by a different counter, therefore, the time need not be specific. When the operation is nearly finished, timing switches back to the cycle timing counter.

A *break* is used to move a word either from an input device to memory or from memory to an output device. The break is only 6 μs in duration but several breaks may run consecutively.

The cycles—program operate A and operate B—are called *machine cycles* and each requires the support of a memory cycle. Program and operate A require a *retrieve* cycle, and operate B requires a *storage* cycle. The memory cycle is not at the exact time as the machine cycle but is rather *staggered* in order to have the information ready at the proper time during the machine cycle. This is shown in Figure 2-14.

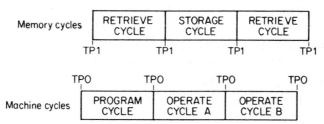

Fig. 2-14. Relation of Memory and Machine Cycles.

Also, the instruction timing does not start and stop with the same TPs that start and stop machine cycles. This is another *time saving* device which allows preparation of a new instruction while the last instruction is being completed. The instruction timing starts at *TP7* of a program cycle. At this time the instruction is out of memory and ready for use. The instruction ends at *TP6* of another program cycle. The next instruction is ready for use and the next timing pulse (TP7) starts timing for that instruction. This is shown in Figure 2-15.

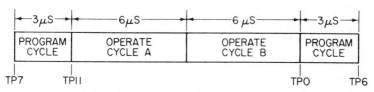

Fig. 2-15. Instruction Timing.

Another important area of the control section is the *program counter*. Its function is to provide the address of the instruction word. Before starting a program, the operator manually inserts the address of the first instruction into the program counter. As long as the program follows a normal sequence, the program provides the address of the next instruction by *adding one* to its count each time an instruction is brought out of memory. When the program *jumps* out of sequence, one of the instructions will place a *new number* in the counter. This obtains the first instruction of a new sequence, and the counter continues to increment by one until another change is required.

In a sketchy fashion, we have covered most of the high points of computer operations. Let's tie them together with a short summary.

2-4 SUMMARY

1. Coded instruction words are stored into the designated program area of memory. Each word consists of an *instruction* and an *address* of the data to be used with that instruction. Each word is stored in a memory register which has a specific address. Normally the instructions are stored into *sequential* locations in the order that they are to be executed.

2. Data words are facts and figures which are stored in *any* available register in the designated data area of memory.

3. The principal area of the arithmetic section is composed of two registers and an adder. The adder can perform mathematical functions using the quantities in the two registers. We have designated one register as A and the other as accumulator.

4. The accumulator contains the *result* at the conclusion of each calculation. The information in the accumulator, and in some cases the A register, can be *manipulated* in several ways. The data in the accumulator can be shifted, compared, altered, transferred to other registers, and sent back to memory storage.

5. All functions are under *program control* and are initiated and controlled by the control section. All timing and control functions are generated in the *control section*.

6. The *master clock* oscillator produces a continuous string of timing pulses. The timing counter *divides* these pulses into groups which represent the time for one cycle. In our sample machine a *cycle* is 6 μs in duration and contains 12 timing pulses separated by 0.5 μs.

7. A memory cycle starts with TP1, and there is a *storage* cycle and a *retrieve* cycle. The storage cycle *destroys* the contents of the selected memory location and then *inserts* the new word. The retrieve cycle *removes* the content of the selected memory location and gates it out of memory but it also *restores* that word into the memory register which it was taken from.

8. There are *three* distinct machine cycles: Program, operate A, and operate B. Each cycle is 6 μs in duration and is supported by a memory cycle. Instructions are removed from memory and decoded during the *program cycle*. Data is obtained from memory and operated on during the *operate cycle A*. A new data word is stored into a selected memory location during the *operate cycle B*. The machine cycles are interspersed and staggered with respect to instruction timing and the memory cycles. This arrangement *saves time* by having information ready for use when it is needed.

9. Most instructions in our machine contain *at least* two machine cycles: Program and either operate A or B. Some instructions have all three cycles. A few instructions require more than 18 μs. In this case a cycle is interrupted, timing is switched to another counter, and the operation is allowed to complete. Interruptions for mathematical operations vary in duration, and they are called *pauses*. Interruptions for moving words into or out of memory are called *breaks*. A break lasts for 6 μs, and several breaks may run in sequence.

10. During the program cycle, the instruction portion of the instruction word is moved from memory into the *operation* register, and the address portion of the word goes into the *address* register. The instruction matrix *decodes* the instruction and *generates* trigger pulses called commands and subcommands which initiate all required actions. The address portion is sent back to memory to select the location which contains the required data.

11. The *program counter* provides the address for each instruction word. The operator manually inserts the address of the *first* instruction. Subsequent instructions are obtained by *adding one* to the count as each instruction is removed. When the program jumps out of sequence, an instruction inserts a new number into the program counter, and the counter continues its sequence from that point.

2-5 FUNCTIONAL BLOCK DIAGRAM

At this point, we should be able to *trace the movement* of instruction words and data words through a functional block diagram. If we can, we have a pretty fair grasp of the *big picture* of computer operations. Oh, there are many details that still need to be fitted in, but our mental picture at this time should at least have the quality of a charcoal sketch.

We will use the instructions that we *designed* earlier. They were:

$$10 = \text{clear and add (CAD)}$$
$$11 = \text{add (ADD)}$$
$$12 = \text{store (STR)}$$
$$13 = \text{halt (HLT)}$$

We will use these to *generate* movement and trace that movement on Figure 2-16.

Instruction Functions

Refer back to Figure 2-4; our program has been stored into sequential locations starting at memory address 1. Our program consists of:

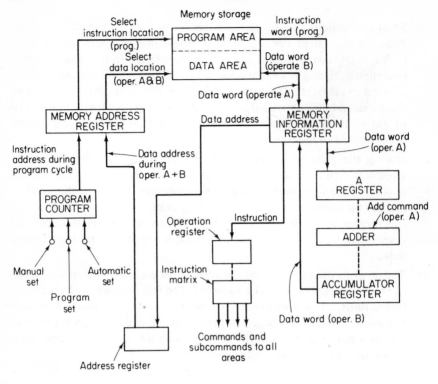

Fig. 2-16. Functional Block Diagram of Central Computer.

Location	Instruction	Data Address
1	10 (CAD)	56
2	11 (ADD)	57
3	12 (STR)	58
4	13 (HLT)	00

CAD: Causes actions which will clear C(Acuum) to zero, and will then add C(56) to the cleared Accum. Back to Figure 2-4, C(56) = 1234. After CAD 56, C(Accum) will be 1234.

ADD: Moves C(57) into the A register and adds it to C(Accum). After ADD 57, C(Accum) = 6904

STR: Clears the contents of memory register 58 and replaces it with C(Accum), 6904. After STR 58, C(58) and C(Accum) = 6904

HLT: Causes automatic operation to cease. The 00 in the address portion of the word has no meaning.

These four instructions will generate all of our *normal* cycles. We will now trace the action on the block diagram (Figure 2-16) for each instruction.

CAD 56

Both program and data are already *stored*. The operator manually sets the *program counter* to a count of 1. This is the memory address of our first instruction word. He then pushes the *start* button.

Since the program is running for the first time, a few μs (3 μs) will be wasted in obtaining the first instruction. The first *program cycle* begins at TP0, but *instruction timing* starts with TP7. The first action is to move the instruction address from the program counter (PC) into the memory address register. The address in the counter is *incremented* by 1. C(PC) = 2, the address of our *next* instruction.

Address 1 from the address register *selects* memory location 1 and a memory *retrieve* cycle starts. During the first half of this cycle, C(1), 1056, moves from memory register 1 into the memory information register (MIR). At this time C(1) = 0000.

During the second half of the retrieve cycle, C(MIR), 1056, is transferred in two directions. This instruction word goes back into the register at memory location 1 and moves out to the operation and address registers.

The instruction portion of the word (10) is *decoded* in the operation register, during the remainder of this program cycle.

TP0 of *operate cycle A* moves the address into the memory address register. TP1 *selects* the specified data location (56) and starts a memory *retrieve* cycle to remove the data which it contains. C(56) are read out to the MIR, and then they are replaced into location 56 and moved into the A register. The contents of location 56, MIR, and the A register are the same (1234).

Next the accumulator is cleared and the *CAD command* causes the 1234 of the A register to be added to the 0000 of the accumulator. When the addition is completed, the accumulator contains the *sum* of the two numbers. In this case, that sum is 1234.

During the last few operations of the CAD instruction, another program cycle is started. The CAD instruction terminates with TP6, and at this time, the *next* instruction word (from location 2) has been moved into the MIR. The PC has been *incremented* to 3; the address of the next instruction.

ADD 57

TP7 of the *same* program cycle that terminated CAD starts the timing for the ADD instruction. This pulse moves the instruction portion of the word (11) from MIR to the *operation* register and the address portion (57) from MIR to the *address* register. The instruction matrix now *decodes* the ADD instruction from the digits 11.

An *operate cycle A* follows the program cycle. The data address (57) moves into the memory address register (MAR) with TP0, and TP1 starts a *retrieve* cycle to remove data from location 57. The data in location 57 (5670) are moved into MIR, rewritten into location 57, and transferred to the A register. The *add command* causes an addition to take place.

The 5670 of the A register adds to the 1234 of the accumulator to produce a sum and new C(Accum) of 6904.

The last few operations of the ADD instruction are still in progress when operate cycle A gives way to another program cycle. C(PC) moves to MAR and selects the instruction word from location 3. This is 1258 (or STR 58). The instruction word moves to MIR and the PC is incremented to 4 just as the ADD instruction terminates.

STR 58

TP7 of the same program cycle which terminated the ADD instruction starts the store instruction. The instruction portion of the word (12) moves to the operation register where it is decoded as STR. The address portion (58) moves into the address register. This terminates the program cycle and starts an operate cycle.

This operate cycle is operate B. The address moves into the MAR and selects location 58. A memory *storage* cycle starts. Meantime, the accumulator contents (6904) have been moved into the MIR. The memory cycle continues to run; location 58 is cleared to 0000, and the 6904 is written from the MIR into the cleared register.

The work is done before operate cycle B terminates, but the instruction timing continues on into another program cycle. The 4 in the PC transfers to MAR, and the PC increments to 5. Location 4 is selected and the instruction word (1300) comes out to MIR. The store instruction ends at TP6 of the program cycle, and the HLT instruction starts with TP7 of the same cycle.

HLT 00

TP7 of the program cycle moves the instruction portion of the word (13) to the operation register and the address portion (00) to the address register. After the program cycle terminates, some checking is done to make sure all important operations have been *completed*. If nothing further is going on, an *operate cycle A* is generated. Normal common operations occur during the cycle, but information is not moved. At TP11 of this cycle, a *subcommand* is generated to stop the machine, but it goes on into a program cycle.

The 5 in the program counter selects and causes C(5) to move into MIR. At TP6 of this program cycle, all actions are terminated. If location 5 contained an instruction word, this word is ready for use in the MIR when the next instruction starts.

If the machine is restarted without manual change, the first timing pulse is TP7 of the same program cycle that terminated operation. The instruction from MIR would then move on out for decoding without any loss of time.

Most registers outside of the memory bank are composed of *volatile* devices. This means that they lose information when the power is removed. For this reason, restarts after power down are the same as for a new program. The operator clears critical registers and enters the address of the first instruction into the program counter before pressing the start button.

Review Exercises

1. Compare the performance of a payroll clerk with that of a computer performing the same task.
2. At what phase of computer development is the list of specifications essential?
3. What is the most logical way to determine the specifications for a new computer?
4. Which section of the computer determines what to do and when to do it?
5. What controls the actions of the section referred to in item 4?
6. What is a register?
7. What type of information is contained in an instruction word?
8. Describe the arrangement of instruction words in memory storage.
9. How does the operator assure that the program execution begins with a specific instruction?
10. How could the output of a properly functioning computer be rendered unreliable?
11. Figure 2-17 is a block diagram of the central computer. Label each

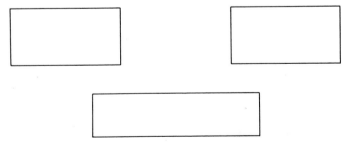

Fig. 2-17. Sections of Central Computer.

section, and draw lines to show control and movement of instructions and data.

12. If a filing drawer is compared to memory storage, what do the numbered folders in the drawer represent?

13. Explain the meaning of this abbreviation: C(500) = 2000.

14. Describe the function of the program counter.

15. Which instruction moves the contents of a memory location into the accumulator? How is this accomplished?

16. Which instruction moves the contents of the accumulator into memory storage? Describe the data path.

17. When the STR instruction is decoded, how does the control section select the proper memory location?

18. Briefly describe the actions that take place during a memory retrieve cycle.

19. Most central computer memories have destructive read out. What measures are taken to avoid loss of information?

20. A stable oscillator is the basis of all computer timing. What do we call this oscillator and its associated circuits?

21. Distinguish between clock pulses and timing pulses.

22. Name the three machine cycles and the primary action that takes place during each of them.

23. Explain why instruction timing and memory cycles are staggered with respect to machine cycles.

computer language

One day computers will use *our* language. They will be able to *speak* as well as they now print information. Figure 3-1 is a picture of a recording produced by Bell Laboratories. On this record the computer speaks and sings quite clearly, and every note is precisely on key.

There will be a time when we talk face to face with a computer, but that time is in the future. The computer of today responds to only one type of language, and that is a language of *numbers*. For the next several years, if we wish to communicate with a computer, we must take the initiative by learning its number language.

3-1 NUMBER SYSTEMS

Many number systems have been devised. A system may be devised by using any set of *symbols* that has a *base* (RADIX) and follows a *logical* progression.

BAND 1 The computer speaking   BAND 2 The computer reciting a soliloquy from Hamlet   BAND 3 The computer singing

Fig. 3-1. Record of a Computer's Voice. (Courtesy Bell Laboratories)

The most familiar system to most of us is the *decimal* system. This system with its radix of ten and ten symbols, 0 through 9, serves us well in most situations. We like this system because it is most familiar to us. If we had grown up with any other system, we would have preferred that system.

Some of the many systems that have been used are:

Binary—base two.

Ternary—base three.

Quaternary—base four.

Quinary—base five.

Senary—base six.

Septenary—base seven.

Octanary—base eight.

Novenary—base nine.

Decimal—base ten.

Unodecimal—base eleven.

Duodecimal—base twelve.

Hexadecimal—base sixteen.

The *number* of different symbols used in a system is the base or *radix* of that system. Figure 3-2 shows the symbols, radix, and logical progression of the systems listed previously.

Four of these twelve number systems have special *significance* to the

Two	Three	Four	Five	Six	Seven	Eight	Nine	Ten	Eleven	Twelve	Sixteen
0	0	0	0	0	0	0	0	0	0	0	0
1	1	1	1	1	1	1	1	1	1	1	1
10	2	2	2	2	2	2	2	2	2	2	2
11	10	3	3	3	3	3	3	3	3	3	3
100	11	10	4	4	4	4	4	4	4	4	4
101	12	11	10	5	5	5	5	5	5	5	5
110	20	12	11	10	6	6	6	6	6	6	6
111	21	13	12	11	10	7	7	7	7	7	7
1000	22	20	13	12	11	10	8	8	8	8	8
1001	100	21	14	13	12	11	10	9	9	9	9
1010	101	22	20	14	13	12	11	10	A	A	A
1011	102	23	21	15	14	13	12	11	10	B	B
1100	110	30	22	20	15	14	13	12	11	10	C
1101	111	31	23	21	16	15	14	13	12	11	D
1110	112	32	24	22	20	16	15	14	13	12	E
1111	120	33	30	23	21	17	16	15	14	13	F
10000	121	100	31	24	22	20	17	16	15	14	10
10001	122	101	32	25	23	21	18	17	16	15	11
10010	200	102	33	30	24	22	20	18	17	16	12
10011	201	103	34	31	25	23	21	19	18	17	13
10100	202	110	40	32	26	24	22	20	19	18	14
10101	210	111	41	33	30	25	23	21	1A	19	15
10110	211	112	42	34	31	26	24	22	20	1A	16
10111	212	113	43	35	32	27	25	23	21	1B	17
11000	220	120	44	40	33	30	26	24	22	20	18
11001	221	121	100	41	34	31	27	25	23	21	19

Fig. 3-2. Several Number Systems.

science of computers. These are the binary system, the octonary system, the decimal system and the hexadecimal system. All of the 12 number systems have a weighted code; the value of a digit is determined by the position it holds. *Positional notation* will describe any number in any number system, and it affords a method of converting a number in one system to an equal quantity in any other system.

3-2　DECIMAL SYSTEM

We seldom take time to analyze a tool that we use every day. So, let's pause for a moment and examine this decimal number system. Some characteristics about this system which we know, but seldom think about, will help us understand the other systems.

Position Values

The decimal system is composed of *ten* symbols which we call digits. They are 0, 1, 2, 3, 4, 5, 6, 7, 8, and 9. With these ten symbols we can express any quantity. This is possible because each number has a weight according to its position with respect to the *decimal point*. This is illustrated in Figure 3-3.

Millions	Hundred thousands	Ten thousands	Thousands	Hundreds	Tens	Units	Decimal points	Tenths	Hundredths	Thousandths	Ten-thousandths	Hundred-thousandths	Millionths
0	0	0	1	6	9	3	.	2	5	3	0	0	0

Fig. 3-3. Decimal Position Value.

The number written in the Figure is 1693.253, and we read it, "one thousand six hundred ninety three *and* two hundred fifty three thousandths. The entire number is simply the *sum of the products* of each digit times its place value: $1000 + 600 + 90 + 3 + 0.2 + 0.05 + 0.003 = 1693.253$. If there is a possibility of confusion as to what number system is used, it can be clarified by writing the radix as a *subscript* to the number, like this 1693.253_{10}.

The position value of a digit may also be expressed as a power of the base.

Number		Powers of 10		Expressed in English
0.000001	=	10^{-6}	=	ten to the negative sixth power
0.00001	=	10^{-5}	=	ten to the negative fifth power
0.0001	=	10^{-4}	=	ten to the negative fourth power
0.001	=	10^{-3}	=	ten to the negative third power
0.01	=	10^{-2}	=	ten to the negative second power
0.1	=	10^{-1}	=	ten to the negative first power
1	=	10^{0}	=	ten to the zero power
10	=	10^{1}	=	ten to the first power
100	=	10^{2}	=	ten to the second power
1000	=	10^{3}	=	ten to the third power
10000	=	10^{4}	=	ten to the fourth power
100000	=	10^{5}	=	ten to the fifth power
1000000	=	10^{6}	=	ten to the sixth power
10000000	=	10^{7}	=	ten to the seventh power

Fig. 3-4. Decimal Place Positions as Powers of Ten.

This is shown in Figure 3-4. The decimal point now *separates* the positive powers from the negative powers. Using our previous number, each digit is worth powers as follows:

$$10^3 \quad 10^2 \quad 10^1 \quad 10^0 \qquad 10^{-1} \quad 10^{-2} \quad 10^{-3}$$
$$1 \quad\; 6 \quad\; 9 \quad\; 3 \qquad\quad 2 \quad\;\; 5 \quad\;\; 3$$

Any number to the *zero power* is equivalent to a one. Again, the value of the number is the sum of the products of each digit times its power of ten. (1 × 10^3) + (6 × 10^2) + (9 × 10^1) + (3 × 10^0) + (2 × 10^{-1}) + (5 × 10^{-2}) + (3 × 10^{-3}) = 1000 + 600 + 90 + 3 + 0.2 + 0.05 + 0.003 = 1693.253.

The digit of any given number which occupies the position of least weight is the *lowest significant digit* (LSD). The digit in the position of the greatest weight is the *most significant digit* (MSD). This might appear too obvious for comment because we always write our numbers with the MSD on the left and the LSD on the right. However, this is a rule, not a law. Inside the computer, a digit is a pulse of voltage or current, and a number of several digits is a string of pulses. In this string of digits the MSD may be on either end. When you observe the output of a counter on an oscilloscope, the pulses appear by time sequence from left to right. In this case the LSD is on the left. When numbers are not written in the conventional fashion, either the LSD or MSD must be specified.

1693.253		352.3961	
MSD	LSD	LSD	MSD

Multiplication and Division by Shifting

Other common operations that we perform *automatically* in the decimal system are multiplication and division by the radix. We do this by shifting the decimal point. When we wish to *multiply* a number by 10, we move the decimal one position to the right: $1693.253 \times 10 = 16932.53$. Mentally we shifted the decimal *point to the right*. The same result is obtained by shifting each digit of the *number* one position *to the left*. In the computer, the point is held and the number is shifted.

When *dividing* by ten, we shift the decimal point to the *left*. The computer shifts the *number to the right*. In either case, $1693.253/10 = 169.3253$.

Relating these operations to the powers of ten, it should be noted that multiplication by 10 *generates* an additional *positive* power of ten while *reducing* the *negative* powers by one. The reverse is true of division.

Application in Computers

The decimal number system is too bulky for efficient use in the digital computer. We would need ten distinctively different conditions to represent the ten symbols. Oh, it could be done. We could use ten different lines and assign a symbol to each line. For ordinary counting this wouldn't be too bad; line number one would always carry zeros, and line number 10 would always carry nines. But how would we represent multiples of a digit? It would require a very complex arrangement of either lines and storage devices or timing circuits. For this reason nearly all digital computers use only binary numbers.

3-3 BINARY SYSTEM

The binary number system is extremely simple, having only the symbols of 0 and 1 (refer to Figure 3-2). Yet, this base two system can be used for *any operation* that involves the use of numbers. Any quantity can be expressed in *binary* that we can express in decimal. It simply requires more binary digits to express a given quantity. For instance, the quantity eight decimal is expressed by the single digit, 8. The same quantity expressed in binary digits is 1000. But since any configuration is composed of only ones and zeros, the circuits which represent these two conditions can be extremely simple.

Simulating the Symbols

Any two extremes can be used to accurately simulate the binary symbols. The highest of two voltage levels commonly represents a one while the lower level is a zero. A closed switch may be a one while the same switch indicates a zero when it is open. Figure 3-5 shows some other examples: Either condition may be used to represent either of the symbols, and you may encounter some confusing variations. However, those outlined in Figure 3-5 are fairly well established.

0	1
Low	High
Off	On
Down	Up
Absent	Present
No hole	Hole
No go	Go
Deenergized	Energized
Nonconducting	Conducting

Fig. 3-5. Conditions Representing Binary Digit Symbols.

Place Values

All of the statements concerning the operations in decimal system are also true for the binary system. Each digit has its own place value and each value can be expressed in the power of the radix (2). Figure 3-6 indicates the decimal equivalent of place values from 2^{27} to 2^{-27}.

Binary Counting

When *counting* in a particular number system, we start with 0 and take each symbol in progressive order until we use all the symbols. Then we carry a 1 to the next higher position and start over with 0 in the first position. In decimal, it is 0 through 9 and then 10, 11, etc.. In binary, we say 0, 1, and we have used all the symbols. The next count carries 1 and starts over with zero. A count of two is 10; three is 11; four is 100; five is 101; six is 110; seven is 111; and eight is 1000.

Figure 3-7 compares the decimal and binary symbol notations of a count from 0 through 15.

2^n	n	2^{-n}
1	0	1.0
2	1	0.5
4	2	0.25
8	3	0.125
16	4	0.062 5
32	5	0.031 25
64	6	0.015 625
128	7	0.007 812 5
256	8	0.003 906 25
512	9	0.001 953 125
1 024	10	0.000 976 562 5
2 048	11	0.000 488 281 25
4 096	12	0.000 244 140 625
8 192	13	0.000 122 070 312 5
16 384	14	0.000 061 035 156 25
32 768	15	0.000 030 517 578 125
65 536	16	0.000 015 258 789 062 5
131 072	17	0.000 007 629 394 531 25
262 144	18	0.000 003 814 697 265 625
524 288	19	0.000 001 907 348 632 812 5
1 048 576	20	0.000 000 953 674 316 406 25
2 097 152	21	0.000 000 476 837 158 203 125
4 194 304	22	0.000 000 238 418 579 101 562 5
8 388 608	23	0.000 000 119 209 289 550 781 25
16 777 216	24	0.000 000 059 604 644 775 390 625
33 554 432	25	0.000 000 029 802 322 387 695 312 5
67 108 864	26	0.000 000 014 901 161 193 847 656 25
134 217 728	27	0.000 000 007 450 580 596 923 828 125

Fig. 3-6. Binary Place Values.

Decimal	Binary	Decimal	Binary
0	0	8	1000
1	1	9	1001
2	10	10	1010
3	11	11	1011
4	100	12	1100
5	101	13	1101
6	110	14	1110
7	111	15	1111

Fig. 3-7. Comparing Decimal and Binary.

Multiplication and Division by Shifting

Multiplication and division can be performed automatically in the binary system by holding the number and moving the *binary point* or by holding the point and moving the number. On paper we would move the point one position to the right to *multiply* by two (the radix). Moving the point one position to the left *divides* the number by two.

In most computers, the binary point is *fixed* except during special operations, but the numbers can be *shifted* left or right. Shifting a number one position to the left is equivalent to multiplying it by two. Shifting a number one position to the right divides it by two.

Converting Binary to Decimal

The quickest way to learn the relationship of two number systems is to convert values between the systems. However, there is a more important reason for learning this conversion. The digital computer deals only with binary numbers, while most of us fully understand only decimal numbers. It is extremely important that we be able to convert the machine's binary to our decimal. Also, when entering information, if we expect the proper response from the computer, we must convert our decimal numbers to the machine's binary notation.

Many methods of converting quantities from one system to another have been devised. Most of them make use of a synthetic form of arithmetic. One of the most reliable and accurate methods has been afforded by the power value of the positions; assign a power of two to each position, multiply each digit by its power, and add the products.

The value of a binary number is equivalent to the sum of the products of each digit times the value of its place position. The number $1111.101_2 = (1 \times 2^4) + (1 \times 2^3) + (1 \times 2^2) + (1 \times 2^1) + (1 \times 2^{-1}) + (0 \times 10^{-2}) + (1 \times 2^{-3}) = 16 + 8 + 4 + 2 + 0.5 + 0.125 = 30.625_{10}$.

Convert this binary number to its decimal equivalent: 101011.101011.

2^5	2^4	2^3	2^2	2^1	2^0		2^{-1}	2^{-2}	2^{-3}	2^{-4}	2^{-5}	2^{-6}
1	0	1	0	1	1		1	0	1	0	1	1

$(1 \times 2^5) + (0 \times 2^4) + (1 \times 2^3) + (0 \times 2^2) + (1 \times 2^1) + (1 \times 2^0) + (1 \times 2^{-1}) + (0 \times 2^{-2}) + (1 \times 2^{-3}) + (0 \times 2^{-4}) + (1 \times 2^{-5}) + (1 \times 2^{-6})$
$= 32 + 0 + 8 + 0 + 2 + 1 + 0.5 + 0 + 0.125 + 0 + 0.03125 + 0.015625 = 43.671875_{10}$.

You may prefer to use another method. Here is one which involves *alternate* multiplication and division. For the *whole number*, the multiplier is 2^1, and for the *fraction*, the multiplier is 2^{-1}. Here are two short examples. Convert 1101 and 0.1011_2 to decimal.

	WHOLE NUMBER	FRACTION
	1 1 0 1	0 . 1 0 1 1
MULTIPLY	$\times$ 2	0.5
	2	0.5
ADD	$+ 1$	1.
	3	1.5
MULTIPLY	$\times$ 2	0.5
	6	0.75
ADD	$+ 0$	0.
	6	0.75
MULTIPLY	$\times$ 2	0.05
	12	0.375
ADD	$+ 1$	1.
	13_{10}	1.375
MULTIPLY		0.5
		0.6875_{10}

The whole number conversion begins with *multiplication* by 2^1 and ends by *adding* the last digit to the product. The fraction conversion starts with *multiplication* by 2^{-1} and ends with a final multiplication.

Let's try a *mixed* binary number with a few more digits, convert 110101.110101_2 to decimal. First *separate* the whole number from the fraction.

	WHOLE NUMBER 110101	FRACTION 0.110101
MULTIPLY	× 2	1
	2	0.5
ADD	+ 1	0.5
	3	0.
MULTIPLY	× 2	0.5
	6	0.5
ADD	+ 0	0.25
	6	1.
MULTIPLY	× 2	1.25
	12	0.5
ADD	+ 1	0.625
	13	0.
MULTIPLY	× 2	0.625
	26	0.5
ADD	+ 0	0.3125
	26	1.
MULTIPLY	× 2	1.3125
	52	0.5
ADD	+ 1	0.65625
	53_{10}	1.
		1.65625
MULTIPLY		0.5
		0.828125_{10}

Thus: $110101_2 = 53_{10}$ and $0.110101_2 = 0.828125_{10}$

When we put the numbers back together we have: $110101.110101_2 = 53.828125_{10}$

Converting Decimal to Binary

The previous process is *reversed* when converting from decimal to binary. First determine the *largest power* of two contained in the number. This provides the place position for the MSD of the binary number; it is always a 1. *Subtract* this power equivalent from the decimal number and move to the next lowest power. If this power is contained in the remainder, the digit is a 1; if not, the digit is a zero. Each time a 1 bit is determined, subtract that power from the remainder. Continue on down to 2^0.

Let's prove this method by converting the 43 of our previous problem back to its binary equivalent.

$$43 > 1 \times 2^5 \quad = \quad 1 \times 2^5 \text{ (MSD)}$$
$$\underline{-32 \quad \text{(Which is } 1 \times 2^5\text{)}}$$
$$11 < 1 \times 2^4 \quad = \quad 0 \times 2^4$$
$$11 > 1 \times 2^3 \quad = \quad 1 \times 2^3$$
$$\underline{-8 \quad \text{(Which is } 1 \times 2^3\text{)}}$$
$$3 < 1 \times 2^2 \quad = \quad 0 \times 2^2$$
$$3 > 1 \times 2^1 \quad = \quad 1 \times 2^1$$
$$\underline{-2 \quad \text{(Which is } 1 \times 2^1\text{)}}$$
$$1 = 1 \times 2^0 \quad = \quad 1 \times 2^0$$
$$\underline{-1 \quad \text{(Which is } 1 \times 2^0\text{)}}$$
$$0$$

When we *collect* the binary digits into the proper power columns, we have 101011_2.

That takes care of the *whole* number; $43_{10} = 101011_2$. Now let's consider the *fraction* 0.671875_{10}. Our highest power and MSD is next to the decimal point, and it is 2^{-1}. It may be either a 1 or a 0. (The symbol $>$ means "is greater than"; $<$ means "is less than.")

$$0.671875 > 1 \times 2^{-1} \quad = \quad 1 \times 2^{-1} \text{ (MSD)}$$
$$\underline{-0.500000 \quad \text{(Which is } 1 \times 2^{-1}\text{)}}$$
$$0.171875 < 1 \times 2^{-2} \quad = \quad 0 \times 2^{-2}$$
$$0.171875 > 1 \times 2^{-3} \quad = \quad 1 \times 2^{-3}$$
$$\underline{-0.125000 \quad \text{(Which is } 1 \times 12^{-3}\text{)}}$$
$$0.046875 < 1 \times 2^{-4} \quad = \quad 0 \times 2^{-4}$$
$$0.046875 > 1 \times 2^{-5} \quad = \quad 1 \times 2^{-5}$$
$$\underline{-0.031250 \quad \text{(Which is } 1 \times 2^{-5}\text{)}}$$
$$0.015625 = 1 \times 2^{-6} \quad = \quad 1 \times 2^{-6}$$
$$\underline{-0.015625 \quad \text{(Which is } 1 \times 2^{-6}\text{)}}$$
$$0$$

Collecting the digits into their indicated place positions produces 0.101011_2. *Combining* this with the whole number conversion indicates that $43.671875_{10} = 101011.101011_2$.

Decimal numbers may be *converted* to binary by applying a *synthetic division* and *multiplication* process. First, the *division method*.

This method is applicable *only to whole numbers*. To translate a whole number (decimal) into binary, *divide* successively by 2 and record the *re-*

mainder in each case. Continue this process until the original number is reduced to 0. The *first* recorded *remainder* is the LSD and the *last* recorded *remainder* is the MSD of the binary number equivalent. Example: Convert 150 decimal into its binary equivalent. Solution:

Divide By 2	Quotient	Remainder	
150/2	75	0	LSD
75/2	37	1	
37/2	18	1	
18/2	9	0	
9/2	4	1	
4/2	2	0	
2/2	1	0	
1/2	0	1-MSD $\rightarrow$ 1 0 0 1 0 1 1 0_2	

The multiplication method is applicable to *fractions*. To change a decimal fraction number into its binary equivalent, *multiply* the fraction successively by 2. Each time the product becomes *greater than 1*, disregard the 1 and continue multiplying the fractional part of the number. For each multiplication that causes the product to become greater than 1—there is a *carry* to the units column—record a 1. For each multiplication that *does not* cause the product to become greater that 1—there is *no carry* to the units column—record a 0. Repeat this procedure until about *three times* as many binary digits as decimal digits are obtained. The *first* recorded bit is the MSD and the *last* recorded bit is the LSD of the equivalent binary fractional number. Example: Convert the decimal fraction 0.225 to its equivalent binary fraction.

Multiply By 2	Product	Carry to Units Column	
0.225 × 2	0.450	0—MSD	
0.450 × 2	0.900	0	
0.900 × 2	1.800	1	
0.800 × 2	1.600	1	
0.600 × 2	1.200	1	0.0 0 1 1 1 0 0 1 1_2
0.200 × 2	0.400	0	
0.400 × 2	0.800	0	
0.800 × 2	1.600	1	
0.600 × 2	1.200	1—LSD	

Note that the first binary digit recorded with either the whole number or the fraction is the digit *next to* the binary point. So, when converting a decimal number to its binary equivalent, start at the binary point and work both ways (to the *left* for the *whole number* and to the *right* for the *fraction*). The LSD and MSD will automatically be correct.

3-4 OCTANARY SYSTEM

The octanary (or octal) number system is based on a radix of *eight*. (Refer back to Figure 3-2.) It contains *eight* symbols which are 0, 1, 2, 3, 4, 5, 6, and 7.

Application to Computers

Octal numbers *are not* used by the computer; neither are they normally used by people. Then why bring in such a number system? The binary system is *difficult* for normal use because of the large number of zeros and ones involved. Because of the large number of digits, the average person is prone to *error* when manipulating binary numbers. Conversion between binary and decimal can be performed, but we need something easier.

The octal number system bridges the gap *between* decimal and binary. We can learn to think in octal notation with a little practice. While learning, the conversion process between octal and decimal is comparatively easy. The beauty of the system, however, lies in the *direct* relationship of each octal symbol to binary notation. Figure 3-8 shows a count of eight expressed in octal and binary.

Octal	Binary
0	000
1	001
2	010
3	011
4	100
5	101
6	110
7	111

Fig. 3-8. Octal and Binary Notations.

This is the *entire* system. Once these eight relationships are learned we can convert *directly* from octal to binary and from binary to octal by visual inspection.

The grouping of three binary digits for each octal digit also serves another useful purpose. Remember that a *computer word* contains a specific number of digits with each binary digit occupying a storage space in a register. Binary digits in the computer are commonly called bits, and we might as well start using that term. By grouping three binary bits to represent each octal digit, it is easy to construct words with the proper number of bits.

Suppose that our computer uses a *word length* of 15 bits and we have this octal number: 123251. Allowing three bits for each symbol produces this binary configuration: 001 010 011 010 101 001. This number has 18 bits; it won't fit into the computer. The octal number can be rounded off by a normal rounding process and reduced to *five* octal digits. In this case we simply drop the LSD. The remaining number, 12325, converts to 001 010 011 010 101 which is a 15 bit word, and it fits nicely into our computer.

Octal Counting

Counting in the octal system is the same as in the decimal system, but we must remember that there are no 8s or 9s. From 0 through 7 the count progresses normally, but 7 + 1 is not 8; it is 10. From 10 through 17, the count is normal, but 17 + 1 is 20. 20 through 27 is normal progression, and 27 + 1 is 30. This continues on and on. Each time 1 is added to a 7 it creates a *carry* into the next higher order column.

Place Values

The *octal point* separates the fraction from the whole number and it separates the positive powers from the negative powers. Since we are dealing with octal numbers, the place values are in terms of *powers of 8*.

POSITIVE POWERS	NEGATIVE POWERS
←	→
1524	4321

Figure 3-9 indicates the decimal equivalent of the powers of eight for nine positions on either side of the octal point.

8^n	n	8^{-n}
1	0	1
8	1	0.125
64	2	0.015 625
512	3	0.001 953 125
4096	4	0.000 244 140 625
32,768	5	0.000 030 517 578 125
262,144	6	0.000 003 814 697 265 625
2,097,152	7	0.000 000 476 837 158 203 125
16,777,216	8	0.000 000 059 604 644 775 390 625
134,217,728	9	0.000 000 007 450 580 596 923 828 125

Fig. 3-9. Powers of Eight.

Octal to Binary and Binary to Octal Conversion

As we have already pointed out, the direct relation between binary and octal removes all the work from conversion. $2^3 = 8^1$, and when we have learned the binary equivalent of the eight octal symbols, we can convert back and forth by simple, direct substitution. Start at the octal point, and work both ways in groups of three. Here they are once more:

Octal	0	1	2	3	4	5	6	7
Binary	000	001	010	011	100	101	110	111

Let's practice on these numbers.

Example 1:
Convert 24615_8 to its binary equivalent.

$$2 \quad 4 \quad 6 \quad 1 \quad 5_8$$
$$010 \quad 100 \quad 110 \quad 001 \quad 101_2$$

Example 2:
Convert 345.26_8 to its binary equivalent.

$$3 \quad 4 \quad 5 \quad . \quad 2 \quad 6_8$$
$$011 \quad 100 \quad 101 \quad . \quad 010 \quad 110_2$$

Example 3:
Convert binary 1011101 to its octal equivalent.

$$001 \quad 011 \quad 101_2$$
$$1 \quad 3 \quad 5_8$$

Example 4:
Convert binary 11101101.11 to its octal equivalent.

$$011 \quad 101 \quad 101 \quad . \quad 110_2$$
$$3 \quad 5 \quad 5 \quad . \quad 6_8$$

Notice that *zeros are added* to make the groups of binary numbers have three digits each.

Octal to Decimal Conversion

Octal to decimal conversion requires a bit more work but becomes very easy after a little practice. Converting directly from the values of the place positions, as we did with binary and decimal, is a reliable method.

$$2 \quad 4 \quad 0 \quad 7 \quad 5 \quad 6 \quad 4 \quad 2_8$$

The decimal equivalent of this octal number is a sum of the products of each digit times its place value. $2(4096) + 4(512) + 0(64) + 7(8) + 5(1) +$

$6(0.125) + 4(0.15625) + 2(0.001953125) = 8192 + 2048 + 0 + 56 + 5 +$
$0.75 + 0.625 + 0.0039 = 10301.3789_{10}.$

Let's try another; convert 623.125_8 to its decimal equivalent.

$$
\begin{array}{llcr}
6 \times 8^2 = 6 \times 64 & & = & 384 \\
2 \times 8^1 = 2 \times 8 & & = & 16 \\
3 \times 8^0 = 3 \times 1 & & = & 3 \\
1 \times 8^{-1} = 1 \times 0.125 & & = & 0.125 \\
2 \times 8^{-2} = 2 \times 0.015625 & & = & 0.03125 \\
5 \times 8^{-3} = 5 \times 0.001953125 & & = & 0.0097656 \\
\hline
& & & 403.1660156_{10}
\end{array}
$$

The repeated *multiplication and addition* method that we used with binary and decimal can also be used here. The procedure is the same. First, we *separate* the whole number from the fraction. We *convert* the whole number and the fraction separately, and then add the parts back together. First we'll try a *whole number*. This results in a repeated progression of *multiplication and addition*. For whole numbers, the multiplier is 8^1, and the addition is the digit in the next lower place position.

Example:

Comvert the octal number 65 to its decimal equivalent.

$$
\begin{array}{cc}
6 & 5 \\
\times\ 8 & \\
\hline
48 & \\
+\ 5 & \\
\hline
53_{10} &
\end{array}
$$

Thus: $65_8 = 53_{10}$

To convert an octal fraction to its decimal fractional equivalent, the multiplier is 8^{-1} or (0.125).

Example:

Convert the octal fraction 0.65 to its decimal fractional equivalent.

$$
\begin{array}{c}
0.6 \quad 5 \\
\times\ \downarrow.125 \\
\hline
+\ 6.625 \\
\times\ 0.125 \\
\hline
33125 \\
13250 \\
6625 \\
\hline
0.828125_{10}
\end{array}
$$

Thus: $0.65_8 = 0.828125_{10}$

Let's try one more example. This time a *mixed number* with a few more digits.

Convert 74205.632_8 to its decimal equivalent. First we'll take the *whole number*.

$$
\begin{array}{ccccc}
7 & 4 & 2 & 0 & 5
\end{array}
$$

$$
\begin{array}{r}
\times\ 8 \\
\hline
56 \\
+\ 4 \\
\hline
60 \\
\times\ 8 \\
\hline
480 \\
+\ 2 \\
\hline
482 \\
\times\ 8 \\
\hline
3856 \\
+\ 0 \\
\hline
3856 \\
\times\ 8 \\
\hline
30848 \\
+\ 5 \\
\hline
30853_{10}
\end{array}
$$

Now, we will convert the *fraction*.

$$
\begin{array}{ccc}
0.6 & 3 & 2
\end{array}
$$

$$
\begin{array}{r}
\times\ 0.125 \\
\hline
3.250 \\
\times\ 0.125 \\
\hline
6.406250 \\
\times\ 0.125 \\
\hline
0.800781250_{10}
\end{array}
$$

Now, we *recombine* the whole number and the fraction, and

$$74205.632_8 = 30853.80078125_{10}$$

Decimal to Octal Conversion

When converting from decimal to octal, it is good to remember that the *place values* provide a straightforward method of conversion between any number systems. We first determine the largest power of eight contained in the decimal number, and this provides the place position of the MSD octal number. *Subtract* the largest number of this power equivalent from the decimal number and move to the next lower power. Let's take the whole number from the previous example for practice: 30853_{10} to octal. Since we just converted from octal, we know that the octal equivalent of this number is 74205.

$$30853_{10}$$
$$- 28672 \quad (7 \times 8^4)$$
$$\overline{\quad 2181 \quad}$$
$$- \quad 2048 \quad (4 \times 8^3)$$
$$\overline{\quad 133 \quad}$$
$$- \quad 128 \quad (2 \times 8^2)$$
$$\overline{\quad 5 \quad}$$
$$- \quad 0 \quad (0 \times 8^1)$$
$$\overline{\quad 5 \quad}$$
$$- \quad 5 \quad (5 \times 8^0)$$
$$\overline{\quad 0 \quad}$$

COLLECT

8^4	8^3	8^2	8^1	8^0
7	4	2	0	5_8

The fractions are handled the same way. The last number that we converted from octal to decimal had a fraction of 0.632_8 which became 0.80078125_{10}. Let's convert it back to octal.

$$0.80078125$$
$$- 0.75 \quad (6 \times 8^{-1})$$
$$\overline{\quad 0.05078125 \quad}$$
$$- 0.046875 \quad (3 \times 8^{-2})$$
$$\overline{\quad 0.00390625 \quad}$$
$$- 0.00390625 \quad (2 \times 8^{-3})$$
$$\overline{\quad 0 \quad}$$

COLLECT

8^{-1}	8^{-2}	8^{-3}
0.6	3	2_8

Another example: Convert 8956.54_{10} to octal.

$$8956.54_{10}$$
$$- 8192 \qquad\qquad (2 \times 8^4)$$
$$\overline{\quad 764.54 \quad}$$
$$- 512 \qquad\qquad (1 \times 8^3)$$
$$\overline{\quad 252.54 \quad}$$
$$- 192 \qquad\qquad (3 \times 8^2)$$
$$\overline{\quad 60.54 \quad}$$
$$- 56 \qquad\qquad (7 \times 8^1)$$
$$\overline{\quad 4.54 \quad}$$
$$- 4 \qquad\qquad (4 \times 8^0)$$
$$\overline{\quad 0.54 \quad}$$
$$- 0.5 \qquad\qquad (4 \times 8^{-1})$$
$$\overline{\quad 0.04000 \quad}$$
$$- 0.03125 \qquad\qquad (2 \times 8^{-2})$$
$$\overline{\quad 0.0087500 \quad}$$
$$- 0.0078125 \qquad\qquad (4 \times 8^{-3})$$
$$\overline{\quad 0.0009375 \quad}$$

We could carry this out further, but let's stop here and collect the digits. We have

8^4	8^3	8^2	8^1	8^0	8^{-1}	8^{-2}	8^{-3}
2	1	3	7	4	4	2	4_8

So, $8956.54_{10} = 21374.424_8$

We can also use the *division and multiplication* process for converting from decimal to octal. First, we *separate* the whole number from the fraction. To convert a *whole number* decimal into its octal equivalent, *divide* successively by 8 and record the remainder in each case. Continue this process until the original number is reduced to 0. The *first* recorded *remainder* is the LSD and the *last* recorded *remainder* is the MSD of the octal number equivalent.

Example:

Translate 150_{10} to its octal equivalent.

Divide by 8	Quotient	Remainder
150/8	18	6—LSD
18/8	2	2
2/8	0	2—MSD $\rightarrow$ 2 2 6_8

To change a decimal fraction into its octal equivalent, *multiply* the fraction successively by 8. Each time the product becomes 1 or greater, disregard the whole number in the product and continue multiplying the fractional part of the number. For each multiplication that causes the product to become 1 or greater, record the whole number. For each multiplication that *does not* result in a whole number product, record a 0. Repeat this process until the octal fraction has *one more* digit than did the decimal fraction.

Example:

Change 0.384_{10} to its octal fractional equivalent.

Multiply by 8	Product	Whole Number
0.384×8	3.072	3—MSD
0.072×8	0.576	0
		0.3 0 4 4_8
0.576×8	4.608	4
0.608×8	4.864	4—LSD

Let's take one more for practice. Convert 3965.267_{10} to octal. Taking the whole number:

$$3965/8 = 495 \text{ remainder } 5$$
$$495/8 = 61 \text{ remainder } 7$$
$$61/8 = 7 \text{ remainder } 5$$
$$7/8 = 0 \text{ remainder } 7 \text{ (MSD)}$$
$$3965_{10} = 7575_8$$

Taking the fraction:

$$0.267 \times 8 = 2.136 \qquad 2 \qquad \text{(MSD)}$$
$$0.136 \times 8 = 1.088 \qquad 1$$
$$0.088 \times 8 = 0.704 \qquad 0$$
$$0.704 \times 8 = 5.632 \qquad 5$$
$$\text{So, } 0.267_{10} = 0.2105_8$$

Putting the number back together, we have $3965.267_{10} = 7575.2105_8$.

3-5 HEXADECIMAL SYSTEM

The hexadecimal number system is based on a *radix of 16*. This means that it must have 16 symbols. Since our decimal system has only *ten* symbols (0 through 9), we will use *six* capital letters from our alphabet for the remaining six number symbols.

Application in Computers

Some computers are designed to transfer information to and from storage in groups of *eight* binary bits. Each group of eight bits is called a *byte*. (Byte is used to designate any common number of bits.) In these machines, the programmer has the *option* of using hexadecimal, binary, or some form of binary coded decimal.

Counting in the Hexadecimal System

In the hexadecimal system, the count proceeds the same as in the decimal system from 0 through 9, then $9 + 1 = A$, $9 + 2 = B$, $9 + 3 = C$, $9 + 4 = D$, $9 + 5 = E$, $9 + 6 = F$, and $9 + 7 = 10$. This completes the cycle and the count appears normal from 10 through 19. When we pass 19, $19 + 1 = 1A$, $19 + 2 = 1B$, $19 + 3 = 1C$, $19 + 4 = 1D$, $19 + 5 = 1E$, $19 + 6 = 1F$, and $19 + 7 = 20$. These should be sufficient examples to show that the count progresses in a normal fashion with a *carry* after each count of 16.

Converting Between Hexadecimal and Decimal

The conversion methods previously described can be applied here, but it is *awkward* at first to perform mathematics with A, B, C, D, E, and F. Therefore, we will resort to *tables*. The values in Figure 3-10 should be sufficient to convert almost any numbers.

With a little help from the table and a little carryover from previous exercises, let's convert $E4B7_{16}$ to decimal. E is to the third power of 16, 4 is to the second power, B is to the first power, and 7 is to the 0 power.

Hexa-decimal	Decimal	Hexa-decimal	Decimal
0	0	1000	4096
1	1	2000	8192
2	2	3000	12288
3	3	4000	16384
4	4	5000	20484
5	5	6000	24576
6	6	7000	28672
7	7	8000	32768
8	8	9000	36864
9	9	A000	40960
A	10	B000	45056
B	11	C000	49152
C	12	D000	53248
D	13	E000	57344
E	14	F000	61440
F	15	10000	65536
10	16	11000	69632
11	17	12000	73728

Fig. 3-10. Hexadecimal-Decimal Table.

$$E4B7 = (E \times 16^3) + (4 \times 16^2) + (B \times 16^1) +$$
$$(7 \times 16^0) = 57344 + 1024 + 176 + 7 = 58551_{10}$$

Relation of Binary and Hexadecimal

When we use four binary digits to express each hexadecimal digit the relationship in Figure 3-11 becomes obvious. Thus, $16DB_{16}$ becomes 0001 0110 1101 1011$_2$. This grouping should clarify that statement:

1	6	D	B
0001	0110	1101	1011

The four digit binary code for hexadecimal expresses the same values as the three digit code for octal.

1	6	D	B_{16}
0001	0110	1101	1011$_2$
1	3 3	3	3$_8$

Hexa–decimal	Binary	Decimal
0	0000	0
1	0001	1
2	0010	2
3	0011	3
4	0100	4
5	0101	5
6	0110	6
7	0111	7
8	1000	8
9	1001	9
A	1010	10
B	1011	11
C	1100	12
D	1101	13
E	1110	14
F	1111	15

Fig. 3-11. Comparison of Systems.

$$5 \quad 7 \quad 2 \quad 3 \quad 4_8$$
$$101 \quad 111 \quad 010 \quad 011 \quad 100_2$$
$$5 \quad E \quad 9 \quad C_{16}$$

This enables *direct* conversion from hexadecimal to binary to octal and from octal to binary to hexadecimal. When converting from hexadecimal to decimal, or vice versa, many people prefer to take the *indirect* route through binary and octal.

3-6 OTHER NUMBER CODES

Several *weighted* codes have been developed for use in the computer. All such codes are intended to bridge the gap between the programmer who *thinks* in decimal and the computer which handles only binary.

8421 Code

The 8421 code is the name of one of the binary coded decimal systems. The 8421 indicates the weight of the binary digit in each of the four columns. We can readily see that 1111_2 with these assigned weights has a decimal equivalent of 15. This represents a certain amount of wasted circuits since the decimal system has only 10 digits, but this code has been widely used and is still used to some extent. Figure 3-12 shows the code for each of the 10 digits.

Decimal	8421 BCD
0	0000
1	0001
2	0010
3	0011
4	0100
5	0101
6	0110
7	0111
8	1000
9	1001

Fig. 3-12. Decimal and 8421 Code.

The 8421 Code is very useful when it is desirable for printouts to be in decimal numbers. But inside the computer the code *is not* flexible. For instance: the *complement* of a number is the remainder when that number is subtracted from the highest symbol in the number system. The complement of: 9 = 0, 8 = 1, 7 = 2, etc..

The computer complements its numbers by *changing* ones to zeros and zeros to ones. The computer complement of 0101 is 1010. When this is converted back to decimal it says that the complement of 5 is 10; it should be 4. This fallacy in the code renders it *useless* for mathematical operations.

Excess Three Code

This code was developed to overcome the *weakness* of the 8421 code. It is constructed by simply adding a binary three to each of the 8421 characters. This produces a code that can be *complemented* by the computer and used mathematically. The codes are compared in Figure 3-13.

Several other codes have been used and new ones are likely to be developed. If you *do not* find a code to your liking, feel free to develop some of your own. None of the codes are very complicated, and we have covered these in sufficient detail to enable you to learn others with little difficulty.

Decimal	8421	Excess 3	Complement
0	0000	0011	1100
1	0001	0100	1011
2	0010	0101	1010
3	0011	0110	1001
4	0100	0111	1000
5	0101	1000	0111
6	0110	1001	0110
7	0111	1010	0101
8	1000	1011	0100
9	1001	1100	0011

Fig. 3-13. Comparison of Codes.

Review Exercises

1. What is the only type of language a computer will respond to?

2. Name the essential ingredients of a number system.

3. What is the radix of each of these number systems:

    (a) Binary?

    (b) Quinary?

    (c) Octanary?

    (d) Hexadecimal?

4. The number 20 represent how many counts in each of these systems:

    (a) Quinary?

    (b) Octanary?

    (c) Hexadecimal?

5. Indicate the weight of each digit in these numbers in terms of powers of the system radix:

    (a) 1010.11_2

    (b) 2064.2_8

    (c) 92.407_{10}

    (d) $A9B.4C_{16}$

6. In the decimal number system, what operation is performed on a number when the decimal point is moved one position to the:

(a) Right?

(b) Left?

7. Inside a computer, what operation is performed by shifting the binary number one place to the:

(a) Right?

(b) Left?

8. Why is the binary number system used almost exclusively inside computers?

9. When each of the following conditions represents a one, what condition would represent zero in each case?

(a) Switch on.

(b) High voltage level.

(c) Hole in a tape.

(d) Transistor conducting.

10. What is the decimal value of 2^{-10}?

11. How many binary bits are required to represent a decimal eight?

12. Convert this number to its decimal equivalent: 1011.101_2.

13. What is the largest power of two contained in each of these decimal numbers:

(a) 55?

(b) 0.65?

(c) 4096?

14. Convert these decimal numbers to their binary equivalents:

(a) 415.24

(b) 4096.4

(c) 1234.12

15. What is the primary use of the octal number system?

16. Convert this octal number to a 15 bit computer word, 3426.

17. What is the decimal equivalent of:

(a) 8^4?

(b) 8^{-4}?

18. Convert 4562.452_8 to decimal

19. Convert 19461.9576_{10} to octal.

20. Convert 5347_8 to binary.

21. Convert 1001101001.0100101_2 to octal.

22. A computer which transfers information in bytes can use which number systems?

23. In the hexadecimal code, how many hexadecimal digits are contained in a byte?

24. Convert the following numbers to binary, octal, and decimal:

    (a) $FOOD_{16}$.

    (b) $ABCDEF_{16}$.

    (c) $EACD_{16}$.

    (d) $18EB_{16}$.

25. Convert the following numbers to octal, binary, and hexadecimal:

    (a) 695468_{10}.

    (b) 57007_{10}.

computer
mathematics

There are many reasons why the computer operator, programmer, and technician should be able to perform basic arithmetic in the *four* number systems previously covered. The operator needs to perform arithmetic from time to time in each number system to *verify* the accuracy of information going into or coming out of the central computer. Of course, he can always convert the numbers to their decimal equivalent, perform the arithmetic in the decimal system, and convert the results back into the original system. We must admit, however, that the operator who can perform arithmetic without conversion is a more skillful operator.

The programmer constantly performs arithmetic operations while *composing* data and anticipating the results of computer operations. Obviously his work is more accurate and more efficient if he is proficient in all four number systems.

It is *doubly important* for the technician to be proficient in all four systems. In addition to performing the function of both operator and programmer from time to time, the technician needs to understand precisely *how the computer performs* each arithmetic operation.

4-1 HEXADECIMAL OPERATIONS

Nearly all digital computer mathematics are limited to the four basic arithmetic operations: *addition, subtraction, multiplication,* and *division.* If the computer is required to extract the square root of a number or compute a set of trigonometry tables, a program *subroutine* will lead it to do so with repeated performance of basic arithmetic. Therefore, we will limit our operations to basic arithmetic.

Mathematics is further simplified in the computer by the fact that it never operates on more than *two numbers* at a time. When it is required to add $2 + 4 + 5 + 3$, it adds $2 + 4$ then adds the sum (6) to the 5, and finally the sum of $6 + 5$ (11) is added to the 3. We will imitate the machine in this respect.

When we reached the stage in grade school that required us to perform multiplication and division, most of us were required to memorize a matrix called a *multiplication table*. While learning the values of the hexadecimal

	1	2	3	4	5	6	7	8	9	A	B	C	D	E	F
1	2	3	4	5	6	7	8	9	A	B	C	D	E	F	10
2	3	4	5	6	7	8	9	A	B	C	D	E	F	10	11
3	4	5	6	7	8	9	A	B	C	D	E	F	10	11	12
4	5	6	7	8	9	A	B	C	D	E	F	10	11	12	13
5	6	7	8	9	A	B	C	D	E	F	10	11	12	13	14
6	7	8	9	A	B	C	D	E	F	10	11	12	13	14	15
7	8	9	A	B	C	D	E	F	10	11	12	13	14	15	16
8	9	A	B	C	D	E	F	10	11	12	13	14	15	16	17
9	A	B	C	D	E	F	10	11	12	13	14	15	16	17	18
A	B	C	D	E	F	10	11	12	13	14	15	16	17	18	19
B	C	D	E	F	10	11	12	13	14	15	16	17	18	19	1A
C	D	E	F	10	11	12	13	14	15	16	17	18	19	1A	1B
D	E	F	10	11	12	13	14	15	16	17	18	19	1A	1B	1C
E	F	10	11	12	13	14	15	16	17	18	19	1A	1B	1C	1D
F	10	11	12	13	14	15	16	17	18	19	1A	1B	1C	1D	1E

Fig. 4-1. Hexadecimal Addition—Subtraction Matrix.

symbols between 9 and 10, that is A, B, C, D, E, and F, we should find such a matrix helpful in hexadecimal arithmetic. Figure 4-1 is a matrix for addition and subtraction, and Figure 4-2 is the hexadecimal multiplication table.

	1	2	3	4	5	6	7	8	9	A	B	C	D	E	F
1	1	2	3	4	5	6	7	8	9	A	B	C	D	E	F
2	2	4	6	8	A	C	E	10	12	14	16	18	1A	1C	1E
3	3	6	9	C	F	12	15	18	1B	1E	21	24	27	2A	2D
4	4	8	C	10	14	18	1C	20	24	28	2C	30	34	38	3C
5	5	A	F	14	19	1E	23	28	2D	32	37	3C	41	46	4B
6	6	C	12	18	1E	24	2A	30	36	3C	42	48	4E	54	5A
7	7	E	15	1C	23	2A	31	38	3F	46	4D	54	5B	62	69
8	8	10	18	20	28	30	38	40	48	50	58	60	68	70	78
9	9	12	1B	24	2D	36	3F	48	51	5A	63	6C	75	7E	87
A	A	14	1E	28	32	3C	46	50	5A	64	6E	78	82	8C	96
B	B	16	21	2C	37	42	4D	58	63	6E	79	84	8F	9A	A5
C	C	18	24	30	3C	48	54	60	6C	78	84	90	9C	A8	B4
D	D	1A	27	34	41	4E	5B	68	75	82	8F	9C	A9	B6	C3
E	E	1C	2A	38	46	54	62	70	7E	8C	9A	A8	B6	C4	D2
F	F	1E	2D	3C	4B	5A	69	78	87	96	A5	B4	C3	D2	E1

Fig. 4-2. Hexadecimal Multiplication—Division Matrix.

Addition

The matrix in Figure 4-1 will help us along. When in doubt as to the *sum* of any two digits, locate one digit on the horizontal margin and the other on the vertical margin. Follow these two points into the matrix to the point of *intersection*. The number at the intersection is the *sum* of the two digits. For example: the sum of D + 9 is 16, and A + F is 19. Until further notice all numbers will be in the hexadecimal number system.

Add:

A	3	7	F	A	3	F	9	8	9	F	9	5	9
6	D	8	B	1	C	5	F	2	D	6	7	E	9
10	10	F	1A	B	F	14	18	A	16	15	10	13	12

When adding more than one column of figures, *carries* are handled exactly as they are in the decimal system, but we must remember that it takes 9 + 7 to equal 10.

Examples:

93	37	86	25	16	AB
FC	C9	BD	93	15	CD
18F	100	143	B8	2B	178

Now let's stretch the numbers a little; a 60 bit computer word can contain the equivalent of 15 hexadecimal digits.

ABCDEF9876	9F8E7D6C5B4A
6789FEDCBA	A4B5C6D7E8F9
11357EE7530	1444444444443

Subtraction

Subtraction is also a very simple process but be careful of the *borrowing* action. We may still use the matrix. This time we locate the subtrahend on either horizontal or vertical margin, follow that line into the matrix to the minuend, then turn 90 degrees and follow a line out to the other margin. The number in this *margin* is the *difference* between the two digits.
Example:

$$D \text{ (Located } within \text{ the matrix)}$$
$$- 8 \text{ (Located on the } left \text{ margin)}$$
$$5 \text{ (Located on the } top \text{ margin)}$$

7	B	F	C	B	E	8	F	E	1E	AB	102
− 3	− A	− 5	− 6	− 4	− 8	− 7	− B	− 5	− F	− 99	− ED
4	1	A	B	7	6	F	4	9	F	12	B6

A63	ABC	1A5	ABCDEF	1234567	A632
− FE	− 8DF	− C9	− 999999	− 89ABCD	− FED
965	1DD	DC	123456	99999A	9645

Multiplication

You may need both matrices for multiplication and division; Figure 4-2 for multiplication and division and Figure 4-1 for addition and subtraction. On the multiplication matrix, we locate one digit on the vertical margin and the other on the horizontal margin. We follow these two points into the matrix. The *product* is located at the *intersection*.
Multiply:

8	6	A	F	C	E	7	9	F
× 5	× 9	× B	× 5	× D	× F	× 7	× 9	× F
28	36	6E	4B	9C	D2	31	51	E1

When multiplying more than one column of digits, the carry is a little tricky, but the procedure is the same as in the decimal number system. For example:

ABC
× F

$$\begin{array}{r} ⒷⒷ \quad \leftarrow \text{CARRIES} \\ A \ B \quad C \\ \times \ F \\ \hline A 1 0 \quad 4 \end{array}$$

Step by step: F × C = B4. The 4 is recorded and the B carries to the next column. F × B = A5. The B carry is added to the 5, producing a 10. The zero from the ten is recorded and the 1 is added to the A to produce the carry, which is another B. F × A = 96. The B carry is added to the 6 to produce 11. One is recorded and one carries to the 9. 9 + 1 = A which is recorded as the final digit.

Multiply:

B4	B4	1234	ABC	6F56
× 3	× D	× 7	× 3	× 2
21C	924	7F6C	2034	DEAC

Using numbers with more digits does not exactly complicate the operation, but it certainly provides more opportunity for *errors* in addition. For example:

24689
× BD

$$\begin{array}{l} ③④⑤⑥ \quad \leftarrow \text{SECOND CARRIES} \\ ③⑤⑥⑦ \quad \leftarrow \text{FIRST CARRIES} \\ 2 \ 4 \ 6 \ 8 \ 9 \end{array}$$

CARRIES	①①① B D
FROM	1D 9 4 F 5
FINAL	190 7 E 3
ADDITION	1AE 1 3 2 5

Let's try another; (ABCD) × (EF).

$$\begin{array}{l} Ⓐ ⒷⒷ \quad \leftarrow \text{SECOND CARRIES} \\ Ⓑ ⒸⒸ \quad \leftarrow \text{FIRST CARRIES} \\ A \ B \ C \ D \end{array}$$

CARRY	① E F
FROM	A 1 1 0 3
FINAL	9 6 5 3 6
ADDITION	A 0 6 4 6 3

Here are some more practice problems on multiplication. We will provide only the problem and the answer; you work out the details.

A9CAC	3B2C	BCAE
× F9	× 4D	× BC
A52634C	11CC3C	8A8FC8

Division

The process here is the same as in any long division operation. When using the matrix in Figure 4-2 for division, we first locate the *divisor* on either margin, follow this point into the matrix to the *dividend*, make a 90° turn and come out to the other margin. The digit in the *margin* at the exit point is the *quotient*. For example:

$$\frac{\text{DIVIDEND}}{\text{DIVISOR}} \quad \frac{A}{5} = 2 \text{ QUOTIENT}$$

We locate the 5 on the first margin (either left or top), follow into the matrix to A, turn and come out to the other margin. At the point of exit, the digit is 2.

Divide:

$$
\begin{array}{ll}
(1) \quad E1/F \quad F \overline{)\begin{array}{l} F \\ E\ 1 \\ \underline{E\ 1} \\ 0 \end{array}}
&
(2) \quad 9A/B \quad B \overline{)\begin{array}{l} E \\ 9\ A \\ \underline{9\ A} \\ 0 \end{array}}
\end{array}
$$

$$
\begin{array}{ll}
(3) \quad 82/D \quad D \overline{)\begin{array}{l} A \\ 8\ 2 \\ \underline{8\ 2} \\ 0 \end{array}}
&
(4) \quad 9C/B \quad B \overline{)\begin{array}{l} E \\ 9\ C \\ \underline{9\ A} \\ 2 \end{array}} \text{REMAINDER}
\end{array}
$$

In problem (4) we have an example of a number that won't divide evenly. Since B will not divide evenly into 9C, we must determine the greatest number times B that is contained in 9C. On the matrix, we find that E × B is 9A. This is less than 9C but the next multiple, F × B = A5, and A5 is greater than 9C. We record E as the quotient, subtract 9A from 9C, and have 2 as a remainder.

We could of course, insert a hexadecimal point after the C, add zeros, and continue the division. Our quotient would then become E.– – – –.

Enlarging the dividend will increase the number of operations, but the problem remains relatively simple.

Examples:

```
            1 2 3 4                    A B C                      6 F 5 6
(1)   7 ⟌ 7 F 6 C      (2)   3 ⟌ 2 0 3 4     (3)   2 ⟌ D E A C
          7                          1 E                        C
        ─────                      ─────                      ─────
          0 F                        2 3                        1 E
          E                          2 1                        1 E
        ─────                      ─────                      ─────
            1 6                        2 4                        0 A
            1 5                        2 4                        A
          ─────                      ─────                      ─────
              1 C                        0                        0 C
              1 C                                                 C
            ─────                                               ─────
                0                                                 0
```

When we enlarge the divisor, the operation remains essentially the same, but we will have to apply more thought to our process. Here is an example: 2B40/B4.

```
            3
   B 4 ⟌ 2 B 4 0
        2 1 C
        ─────
          9 8
```

B4 is larger than 2B, so, we include the next digit. $2 \times B4 = 168$; that's too small. $3 \times B4 = 21C$. C from 14 is 8 and 1 from A is 9.

```
            3 D
   B 4 ⟌ 2 B 4 0
        2 1 C
        ─────
          9 8 0
          9 2 4
        ─────
            5 C
```

Bringing down the 0, we now divide 980 by B4. $C \times B4 = 840$; that's too small. $D \times B4 = 924$. $980 - 924 = 5C$.

```
            3 D. 8
   B 4 ⟌ 2 B 4 0. 0
        2 1 C
        ─────
          9 8 0
          9 2 4
        ─────
            5 C 0
            5 A 0
          ─────
              2 0
```

Adding the hexadecimal point, we can add zeros and continue the division. We now divide 5C0 by B4. $7 \times B4 = 4EC$; that's too small. $8 \times B4 = 5A0$. $5C0 - 5A0 = 20$.

REMAINDER

Here are some practice problems with answers. Work them out and fill in the details.

```
         D                 F                 B                   B O C
(1)  B ⟌ 8 F    (2)  C ⟌ B 4    (3)  F ⟌ A 5    (4)  A ⟌ 6 E 7 8
```

$$
\begin{array}{r}
1\ 2\ D\ B +3 \\
(5)\quad 9\ \overline{\big|\ A\ 9\ B\ 6\ }
\end{array}
\qquad
\begin{array}{r}
7\ C\ A \\
(6)\quad D\ \overline{\big|\ 6\ 5\ 4\ 2\ }
\end{array}
\qquad
\begin{array}{r}
5\ A\ B \\
(7)\quad 2\ B\ \overline{\big|\ F\ 3\ B\ 9\ }
\end{array}
$$

$$
\begin{array}{r}
B\ A\ D \\
(8)\quad 4\ A\ \overline{\big|\ 3\ 6\ 0\ 0\ 2\ }
\end{array}
\qquad
\begin{array}{r}
D\ A\ B \\
(9)\quad D\ 2\ \overline{\big|\ B\ 3\ 6\ 4\ 6\ }
\end{array}
$$

4-3 OCTAL OPERATIONS

Now that we are becoming accustomed to thinking in numbers *other than decimal*, octal arithmetic should be comparatively simple. We could probably handle this number system without the benefit of matrices, but we'll include them in case they are needed. Figure 4-3 is used for addition and subtraction while Figure 4-4 will aid us in multiplication and division.

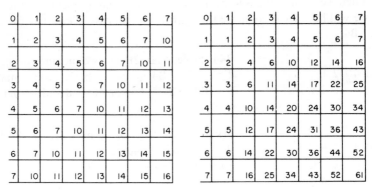

Fig. 4-3. Octal Addition-Subtraction Matrix.   **Fig. 4-4.** Octal Multiplication-Division Matrix.

These matrices function exactly as those we used for hexadecimal, therefore, further explanation is unnecessary.

There is one important point that we must keep in mind when dealing with octal numbers; the system uses *only* 0, 1, 2, 3, 4, 5, 6, and 7; *there are no 8s or 9s.* Yes, we stated that before, and it seems too obvious to need repetition. However, most octal arithmetic errors are caused by momentarily forgetting this fact.

Addition

$$
\begin{array}{llllll}
(1)\ \ 7 & (2)\ \ 5 & (3)\ \ 6 & (4)\ \ 7 & (5)\ \ 4 & (6)\ \ 2 \\
\ \ \ \ \ \underline{1} & \ \ \ \ \ \underline{3} & \ \ \ \ \ \underline{4} & \ \ \ \ \ \underline{6} & \ \ \ \ \ \underline{4} & \ \ \ \ \ \underline{7} \\
\ \ \ 10 & \ \ \ 10 & \ \ \ 12 & \ \ \ 15 & \ \ \ 10 & \ \ \ 11
\end{array}
$$

Did you have *second thoughts* about any of these examples? If not, you are learning to think octally. If you did, just keed reminding yourself that $7 + 1 = 10$.

When more than one column of digits are involved we use the same *sum and carry* technique that we use with decimal numbers. A carry to the next higher order column is produced each time a sum of any two digits in a column equals or exceeds the *radix* of the particular number system.
Example:

Take the columns right to left

$$\begin{array}{r} 447 \\ + 652 \\ \hline 1321 \end{array}$$
$2 + 7 = 11$, bring 1 down and carry 1.
$5 + 4 + 1 = 12$, bring 2 down and carry 1.
$6 + 4 + 1 = 13$, bring 3 down and carry 1.
$0 + 1 = 1$, bring it down. There is no carry.

Check the addition in the following problems.

(1)		(2)		(3)		(4)		(5)	
	7654		641		1234		5432		1234567
	+ 1426		+ 234		+ 4321		+ 3254		+ 7273747
	11302		1075		5555		10706		10530536

Subtraction

The rules for octal subtraction are the same as with any other number system.
Example:

$$\begin{array}{r} 54.3 \\ - 45.4 \\ \hline 6.7 \end{array}$$

Since the 4 is larger than the 3, we borrow 1 from the next higher column. This changes the 3 to 13, and $13 - 4 = 7$. The 4 which we borrowed from is now a 3, and another borrow is in order. Again we have 13 and $13 - 5 = 6$. So, $54.3 - 45.4 = 6.7$.

Prove the operation by adding the difference back to the subtrahend. The operation should produce the minuend as a sum.

$$\begin{array}{r} 45.4 \\ + 6.7 \\ \hline 54.3 \end{array}$$

Here are some practice problems; *verify the answer* in each case.

(1)		(2)		(3)		(4)		(5)	
	477		725		7432		7776		76543210
	− 352		− 647		− 2347		− 6777		− 01234567
	125		56		5063		777		75306421

(6)
$$\begin{array}{r} 7575757575 \\ - 5757575757 \\ \hline 1616161616 \end{array}$$

Another point that bears repeating is the fact that there are no octal numbers *inside* the computer. However, they are used extensively in programming. As a result, much of the computer *input* is in octal form, and a great many output devices supply octal *output*.

Complementing Numbers

Most computers can produce the results of addition, subtraction, multiplication, and division, but they generally do all this by *addition alone*. If we multiply $5 \times 6_{(10)}$, we get a product of $30_{(10)}$. We can obtain the same result by adding $5_{(10)}$ to itself six times: $5 + 5 + 5 + 5 + 5 + 5 = 30$. When we divide $20/4_{(10)}$, we obtain a quotient of 5. The same result can be produced by subtracting 4 from $20_{(10)}$ five times: $20 - 4 = 16 - 4 = 12 - 4 = 8 - 4 = 4 - 4 = 0$.

In the first case, we *multiplied by addition*. In the second case, we *divided by subtracting*. If we can find a way to *subtract by adding*, then all four arithmetic functions can be performed by simple *addition*. That is one of the big reasons why number *complements* are important to a study of computers.

There are *two types* of complements in each number system: the *radix* complement and the *radix-1* complement.

The radix complement is simply the difference between the *digit and the radix* of the number system.

Examples:

	Decimal	Octal	Binary
DIGITS	0279	047	1010
COMPLEMENT	10831	841	1212

Here we considered the digits individually. When they are grouped to form larger numbers, the complement comes out slightly different. The *radix-1 complement* is formed by subtracting each *digit* from the *largest* symbol in the number system. It can be converted to the radix complement by simply *adding a 1* to the least significant digit of the number. Decimal examples:

We will form complements of the numbers 25, 136, and 9874.

LARGEST SYMBOL	99	999	9999
SUBTRACT	25	136	9874
RADIX-1 COMP.	74	863	0125
ADD 1 TO LSD.	1	1	1
RADIX COMP.	75	864	0126

Since we are dealing with decimal numbers at the moment, we call the radix complement a *tens* complement. The radix-1 complement then becomes the *nines* complement. Remember that the tens complement is simply the nines complement *plus one*.

Octal examples:

We will form complements of these numbers: 45, 142, and 7460.

LARGEST SYMBOL	77	777	7777
SUBTRACT	45	142	7460
RADIX-1 COMP.	32	635	0317
ADD 1 TO LSD	1	1	1
RADIX COMP.	33	636	0320

In the octal number system we refer to the radix complement as the *eights* complement and the radix-1 complement becomes the *sevens* complement. Again, please note that the eights complement is simply the sevens complement *plus one*. Binary examples:

We will form complements of these numbers: 01, 101, and 1010.

LARGEST SYMBOL	11	111	1111
SUBTRACT	01	101	1010
RADIX-1 COMP.	10	010	0101
ADD 1 TO LSD	1	1	1
RADIX COMP.	11	011	0110

In the binary system, we refer to the radix complement as the *twos* complement, and the radix-1 complement is the *ones* complement. The twos complement is the ones complement *plus one*.

Now let's get back to our octal arithmetic. Our next step required that we understand octal number *complements*. While on the subject, it seemed best to cover complements in the *three* basic systems at the same time.

Octal Subtraction by Addition

Suppose that we wish to subtract 3542_8 from 7460_8. In normal subtraction, we produce a difference of 3716; like this:

$$
\begin{array}{r}
7460 \\
- 3542 \\
\hline
3716
\end{array}
$$

If we change the subtrahend to the *eights* complement, we can obtain the same result by *adding*.

LARGEST SYMBOL	7777
SUBTRACT	3542
SEVENS COMP.	4235
PLUS 1	1
EIGHTS COMP.	4236
ORIGINAL MINUEND	7460
ADD EIGHTS COMP.	+ 4236
	1 3716
DISCARD END CARRY	

Another popular method of handling complements is to first subtract the sevens complement. We then use the *end carry* (which always occurs) to add to the *LSD* of the result. This corrects the operation and avoids discarding the end carry. Using the same figures as in the previous example, we have:

$$
\begin{array}{r}
7460 \\
+\ 4235 \\
\hline
1\quad 3715 \\
1 \\
\hline
3716
\end{array}
$$

(SEVENS COMP.)

(ADD END CARRY TO LSD)

One more vital point on the mechanics of complementing; both of our numbers must contain the *same number of digits*. This is no problem inside the computer. Remember that the number of storage devices in a register dictates the number of digits in a computer word. And it is physically impossible to have *nothing* in a storage device. A register may be filled with zeros, but that is different from nothing. Suppose we have 15 bit registers and these octal numbers are to be stored: 77777, 526, and 42. Figure 4-5 illustrates the storage of these numbers.

Octal number	To binary	Storage in register
77777	III III III III III	1 1 1 1 1 1 1 1 1 1 1 1 1 1 1
526	IOI OIO IIO	0 0 0 0 0 0 1 0 1 0 1 0 1 1 0
42	IOO OIO	0 0 0 0 0 0 0 0 0 1 0 0 0 1 0

Fig. 4-5. All Numbers in a Given Computer Have the Same Number of Digits.

Our arithmetic should be *compatible* with the computer arithmetic, so let's start using numbers with the same number of digits in both rows. The 526 from Figure 4-5 then becomes 00526 and 42 becomes 00042. Now, when we complement a number to the sevens complement, each 0 becomes a 7 and each 7 becomes a 0. Inside the computer, the same thing is accomplished by simply *reversing the order* of ones and zeros; 1010 becomes 0101. In Figure 4-5, when we complement the 42, it becomes 111 111 111 011 101. Let's convert that to octal and recomplement it.

111	111	111	011	101	ONES COMP.
7	7	7	3	5	SEVENS COMP.
0	0	0	4	2	ORIGINAL NUMBER

Here are some more octal subtraction problems for practice. The answers are provided, and you are expected to *verify the answers*.

Procedure:

1. Obtain the *sevens complement* of the subtrahend.
2. *Add* the complement to the minuend.
3. Add the *end carry* to the LSD.

Problems:

(1)	6542	(2)	77777	(3)	7654321	(4)	00563
	− 0765		− 00000		− 1234567		− 00075
	5555		77777		6417532		00466

(5)	52525	(6)	70707
	− 25252		− 07070
	25253		61617

Octal Multiplication

Unless you have already memorized the octal multiplication table, you will find a need to constantly refer to the matrix in Figure 4-4. Let's turn there now, and please note that $5 \times 5 = 31$ (not 25) and $7 \times 7 = 61$ (not 49).

Multiply:

(1)	7	(2)	4	(3)	6	(4)	5	(5)	7	(6)	3
	× 5		× 3		× 6		× 7		× 2		× 6
	43		14		44		43		16		22

(7)	25	(8)	61	(9)	43	(10)	52	(11)	527
	× 05		× 07		× 06		× 04		× 042
	151		527		322		250		1256
									2534
									26616

(12)	746	(13)	75432
	× 057		× 75432
	6512		173064
	4576		270516
	54472		366150
			463602
			656666
			7314267244

Here are more problems with the answers given. Solve the problems and *verify the answers*.

(14)	432	(15)	7070	(16)	25252
	× 765		× 0707		× 52525
	422342		6242610		1616116162

Octal Division

We will still need the matrix in Figure 4-4 for these operations. Divide:

$$
\begin{array}{r} 3 \\ 07\overline{)25} \\ 25 \\ \hline 0 \end{array}
\qquad
\begin{array}{r} 7 \\ 06\overline{)52} \\ 52 \\ \hline 0 \end{array}
\qquad
\begin{array}{r} 6 \\ 05\overline{)36} \\ 36 \\ \hline 0 \end{array}
\qquad
\begin{array}{r} 2 \\ 25\overline{)52} \\ 52 \\ \hline 0 \end{array}
$$

(1)   (2)   (3)   (4)

$$
\begin{array}{r} 7 \\ 012\overline{)106} \\ 106 \\ \hline 0 \end{array}
\qquad
\begin{array}{r} 6 \\ 036\overline{)264} \\ 264 \\ \hline 0 \end{array}
$$

(5)   (6)

Now some practice problems with answers; you *fill in the details.*

(7) 26616/00527 = 42
(8) 54472/00057 = 746
(9) 122306/003452 = 27
(10) 1705324/0000314 = 4567

4-3 BINARY OPERATIONS

It is *important* to understand binary arithmetic because this is the computer's own language. But it is so *simple* that we will devote only limited space to it.

Binary Addition

Straight addition of binary numbers is not very complicated. About the hardest problem we encounter is $1 + 1 + 1$. Figure 4-6 is a truth table for binary addition.

0	+	0	=	0	&	Carry of 0
0	+	1	=	1	&	Carry of 0
1	+	0	=	1	&	Carry of 0
1	+	1	=	1	&	Carry of 1

Fig. 4-6. Binary Addition.

The table in Figure 4-6 shows the results of *all* possible two digit combinations.

Add:

| (1) | 1
0
$\overline{1}$ | (2) | 0
1
$\overline{1}$ | (3) | 0
0
$\overline{0}$ | (4) | 1
1
$\overline{10}$ | (5) | 11
00
$\overline{11}$ | (6) | 10
01
$\overline{11}$ |

| (7) | 11
01
$\overline{100}$ | (8) | 11
10
$\overline{101}$ | (9) | 11
11
$\overline{110}$ | (10) | 101
010
$\overline{111}$ | (11) | 1111
1111
$\overline{11110}$ |

(12)   101110111101111
     101110000111010
     $\overline{1011101000101001}$

Binary Subtraction

Figure 4-7 is a truth table for subtraction.

0	−	0	=	0	No borrow
0	−	1	=	1	Borrow 1
1	−	0	=	1	No borrow
1	−	1	=	0	No borrow

0	x	0	=	0
0	x	1	=	0
1	x	0	=	0
1	x	1	=	1

Fig. 4-7. Binary Subtraction. **Fig. 4-8.** Binary Multiplication.

Subtract:

| (1) | 1
− 1
$\overline{0}$ | (2) | 0
− 0
$\overline{0}$ | (3) | 10
− 01
$\overline{01}$ | (4) | 11
− 1
$\overline{10}$ | (5) | 111
− 010
$\overline{101}$ | (6) | 111
− 110
$\overline{001}$ |

| (7) | 110
− 011
$\overline{011}$ | (8) | 101010
− 011111
$\overline{001011}$ | (9) | 111110101
− 110101100
$\overline{001001001}$ |

Binary Multiplication

Figure 4-8 is a binary multiplication truth table.

Multiply:

| (1) | 1
× 1
$\overline{1}$ | (2) | 0
× 1
$\overline{0}$ | (3) | 0
× 0
$\overline{0}$ | (4) | 1
× 0
$\overline{0}$ | (5) | 11
× 01
$\overline{11}$ |

(6)   1010   (7)   111110
      × 0101   × 101101
    ─────────   ──────────
        1010   111110
        0000   000000
        1010   111110
        0000   111110
     ─────────   000000
      0110010   111110
                               ──────────
                               101001100110

Binary Division

Division by zero has never been defined, so there is no truth table for this operation. However, the procedure is the same as in any other division process.

Examples:

(1) $\dfrac{111}{101 \overline{)100011}}$
       101
      ─────
        111
        101
       ─────
        101
        101
       ─────
          0

(2) $\dfrac{11}{011 \overline{)001001}}$
       011
      ─────
        011
        011
       ─────
          0

Divide:

(1) 001100/000010 = 110   (2) 011001/000101 = 101
(3) 101010/000110 = 111   (4) 011000/000100 = 110

4-4 MACHINE ARITHMETIC

This is an important *extension* of binary arithmetic. The operations which we perform here in a mechanical fashion will provide considerable insight into actual *computer functions*.

Fixed Point

We have stated that the registers in the central computer all have the same *bit capacity*, and this makes all our numbers (words) the same length. Most arithmetic is performed with the assumption that the *binary point* always appears in the same position. For convenience, the programmer makes a decision to use *either all whole numbers or all fractions*. If he decides to use fractions, the binary point is considered to be to the *left* of the MSD. If he decides to use whole numbers, the binary point is visualized as being to the *right* of the LSD. Figure 4-9 illustrates a register with each of these conditions.

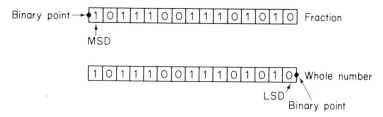

Fig. 4-9. Position of Binary Point.

In the first case, the register contains *a fraction* which converts to 0.56352_8. The second register contains the same digits as a *whole number*, 56352_8. It makes no difference in the operation whether we use fractions or whole numbers, but we should be *consistent*. We can scale our numbers to fit either way, but if we deal with whole numbers part of the time and fractions part of the time, we soon *lose track* of what the machine contains. When this happens, the inputs and outputs become *meaningless* to us.

Floating Point

Some computers are designed with logarithmic capabilities. This means that numbers can be stored and operated on in terms of a *mantissa* and a *characteristic*. Or if you prefer, a *base* number with an *exponent*. Since we are dealing with binary numbers, the characteristic (exponent) will be in powers of two.

Certain bit positions in the registers are considered as the mantissa and the remaining bits form the exponent in powers of two. This is illustrated in Figure 4-10.

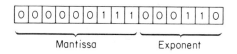

Fig. 4-10. Register With a Floating Point Number.

This register contains a binary 111 raised to the sixth power of two. The computer with this capability can also perform normal fixed point operations. When floating point operations are desired, *floating point instructions* are used and the addresses refer to locations which contain *data* in floating point format.

Floating point numbers are manipulated in the computer in much the same way that we handle the powers of ten. When *addition* is to be done, the two numbers are arranged so that they have the *same* power of two. The mantissas are then *added* and the *common* power is assigned to the new mantissa. The same general procedure is observed in subtraction.

When *multiplying*, the mantissas are *multiplied* and the powers are *added*. In *division*, the mantissas are *divided* and the powers are *subtracted*.

If we designate m as the *mantissa* and x as the *exponent*, we can establish the following rules. The subscripts 1 and 2 refer to the *two* computer words being operated on. We can then represent register contents as shown in Figure 4-11.

Fig. 4-11. Representation of Floating Point Words.

The Processes are:

1. Addition: $(m_1 + m_2)(x_1 \text{ or } x_2)$
2. Subtraction: $(m_1 - m_2)(x_1 \text{ or } x_2)$
3. Multiplication: $(m_1 \times m_2)(x_1 + x_2)$
4. Division: $(m_1/m_2)(x_1 - x_2)$

Values may now be *substituted* for m_1, m_2, x_1, and x_2 to demonstrate the processes. For instance: let's perform *addition* with these values: $m_1 = 111, m_2 = 101, x_1 = 011$, and $x_3 = 011$.

Binary: $(111 + 101)(011) = 001100 \times 011$
Octal: $(7 + 5)(2^3) = 14 \times 2^3$

Using the same values and *multiplying*, we have:

Binary: $(111 \times 101)(011 + 011) = 100011 \times 110$
Octal: $(7 \times 5)(2^3 + {}^3) = 43 \times 2^6$

Negative Numbers

Some computers are designed to handle both positive and negative numbers in their *true form*. Others are designed to handle *negative* numbers in the *one's complement* form. In either case, the computer must have a means of *distinguishing* between positive and negative numbers. The most common method is to assign an *extra* binary bit to each computer word and designate that bit as a *sign* bit. Usually a *zero* sign bit identifies the word as a positive number while a *one* sign bit shows that the number is negative. The sign bit is generally to the left of the MSD register position as shown in Figure 4-12. This is our 15 bit register with an extra bit position added to the left of the MSD position.

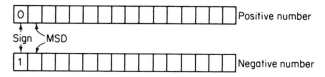

Fig. 4-12. Use of Sign Bit.

If the machine is designed to have all negative numbers in complement form, the input device will *change* the zeros to ones and the ones to zeros as the negative number is transferred into the central computer. In this case, we need to define a *negative zero*. If a word is composed of *all ones*, including the sign bit, the number is a negative zero. On the other hand, when we enter a negative number that does consist of all ones, the sign bit *only* becomes a one and all other bits are complemented to zeros.

When *adding* a positive and a negative number, we take the *difference* and affix the sign of the larger. The computer with the negative numbers in complement form handles it like this:

(1)   SIGN

         0 101000 $(+ 50_8)$

    ADD

         1 111111 $(- 00_8)$

    1 0 100111 $(+ 50_8)$ ADD

               1

        0 101000

   SIGN ∨ ∨ ∨

        + 5 0

(2)   SIGN

         0 111 101 $(+ 75_8)$

    ADD

         1 101 011 $(- 24_8)$ ADD

   1 0 101 000 (51_8)

               1

        0 101 001

        ∨ ∨ ∨

        + 5 1

When we *subtract* a positive *and* a negative number, we change the sign of the subtrahend and *add*. The computer complements *all bits* of the subtrahend and adds.

(1)   SIGN

0 011 $(+3_8)$

SUB

1 101 (-2_8)

COMP. $(+5_8)$

0 011

ADD

0 010

0 101

V V

+ 5

(2)   0 100 010 $(+42_8)$

SUB

1 100 110 (-31_8) SUB

(73_8)

0 100 010

0 011 001

0 111 011

V V V

+ 7 3

When we *multiply* a positive *and* a negative number, our product is always *negative*. The computer first *complements* all bits in the negative word, *multiplies* the two words together, and then *recomplements* the product.

SIGN

1 100 (-3_8)

0 011 $(+3_8)$ MUL

(-11_8)

COMP.

0 011

MUL   0 011

0011

0011

0000

0000

0001 001

COMP.

1 110 110

V V V   (a 6 is a 1 in complement form)

− 1 1

You are undoubtedly aware of the fact that when two binary numbers are multiplied together, the product has *twice* as many digits as either of the numbers. This is provided for by joining *another* full length register to the accumulator during the multiplication process. Since there is no provision for storing a double length word, the product is *rounded off* to a normal word length before storing. The register is shown in Figure 4-13.

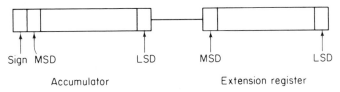

Fig. 4-13. Accumulator Extension.

The rounding off process is very simple. The computer simply adds the *MSD* of the extension register to the *LSD* of the accumulator.

When we *multiply* two numbers of *like signs*, either two positive numbers or two negative numbers, we obtain a *positive* product. The computer handles it like this: *Two positive* numbers—just straight multiplication.

$$
\begin{array}{ccc}
 & 0\ 111 & (+7_8) \\
\text{MUL} & 0\ 111 & (+7_8) \quad \text{MUL} \\
\hline
 & 0\ 111 & (61_8) \\
 & 01\ 11 & \\
 & 011\ 1 & \\
 & 0000 & \\
\hline
 & 0\ 110\ 001 & \\
 & \lor\ \lor\ \ \lor & \\
 & +\ 6\ \ \ 1 &
\end{array}
$$

Two negative numbers—both numbers are first complemented, and then, straight multiplication.

$$
\begin{array}{lcc}
\text{SIGN} & & \\
 & 1\ \ 101 & (-2_8) \\
 & 1\ \ 101 & (-2_8) \quad \text{MUL} \\
\cline{2-2}
\text{COMP} & & (+4_8) \\
 & 0010 & \\
\text{MUL} & 0010 & \\
\cline{2-2}
 & 0000 & \\
 & 0010 & \\
 & 0000 & \\
 & 0000 & \\
\cline{2-2}
 & 0\ 000\ 100 & \\
 & \lor\ \lor\ \ \ \lor & \\
 & +\ 0\ \ \ \ 4 &
\end{array}
$$

When we *divide* numbers of *like signs* the quotient is *positive*. When we divide numbers of *unlike signs* the quotient is *negative*. The computer divides two positive numbers without change. If both numbers are negative, it complements the negative numbers, divides, and keeps the quotient *as is*. When *one number is positive* and the other is negative, the negative number is complemented *before* the division, and the quotient is recomplemented *after* the division is completed.

Two *positive* numbers:

$$
\begin{array}{r}
0\ 000010 \\
0\ 010\overline{)0\ 000100} \\
0\ 00010 \\
\hline
00
\end{array}
\qquad (+4_8/+2_8 = +2_8)
$$

$$
\begin{array}{c}
0\ 000\ 010 \\
\vee\ \ \vee\ \ \vee \\
+\ \ 0\ \ \ 2
\end{array}
$$

Two *negative* numbers:

$$
\begin{array}{l}
1\ 111011/1\ 101 \quad\text{(COMP BOTH)} \qquad (-4_8/-2_8 = +2_8) \\
0\ 010\overline{)0\ 000100}
\end{array}
$$

Continue same as with two positive numbers.

One *positive* and one negative:

$$
1\ 111011/0\ 010 \qquad\qquad (-4_8/+2_8 = -2_8)
$$

$$
\text{(COMP NEG)}
$$

$$
\begin{array}{r}
0\ 000010 \\
0\ 010\overline{)0\ 000100}
\end{array}
$$

COMP QUOTIENT

$$
\begin{array}{c}
1\ 111\ 101 \\
\vee\ \ \vee\ \ \vee \\
-\ \ 0\ \ \ 2
\end{array}
$$

Review Exercises

1. Each part of this item is a *number* representing a small segment of *data*. Convert each number into a *complete* computer word for a *fixed point, whole number, binary* computer. This computer stores negative numbers in *complement* form and a *register* holds 15 significant bits plus a sign bit. A one sign bit represents a *negative*.

a. $+ABC_{16}$ b. $-4F5_{16}$ c. $+1954_{10}$

d. -63_{10} e. $+74523_8$ f. -00000_8

g. $+10101_2$ h. -10101_2 i. -111_2

2. Perform the indicated operation on the following *hexadecimal* numbers.

a. ABCD	b. 75FD	c. FOOD	d. 8B969
+4 5 6 7	− 64DF	× 42	D

3. Add the following *octal* numbers.

a. 7345	b. 63275	c. 57136
+ 5472	+ 25713	+ 71377

4. Perform the indicated subtraction on the following *octal* numbers by the direct method.

a. 73642	b. 51237	c. 54327
− 5764	− 42653	− 37177

5. Perform the indicated subtraction on the following *octal* numbers using the 7's complement and end carry.

a. 35003	b. 32135	c. 40600
− 7526	− 7753	− 37652

6. Multiply these *octal* numbers.

a. 2760	b. 2175	c. 5713
× 453	× 764	× 347

7. Perform the indicated *octal* division carrying the quotient to two octal places.

a. 7306 / 75	b. 4570 / 15	c. 26437 / 437

Items 8 through 13 concern binary numbers.

8. Add:

a. 1101101	b. 11011	c. 0100111
+ 1011010	+ 01010	+ 1011101

9. Subtract using *direct* method:

a. 110101	b. 101101	c. 1110101
− 1011	− 10111	− 111111

10. Subtract using *one's complement* and end carry:

a. 1011101	b. 110110	c. 1001001
− 100110	− 10111	− 101110

11. Multiply:

a. 11111	b. 1011	c. 100010
× 11	× 101	× 111

12. Divide, carrying quotient to *two* binary places.

a. $\dfrac{101101}{110}$ b. $\dfrac{101001}{101}$ c. $\dfrac{110100}{100}$

13. Perform the indicated operations and express the answer in *full computer words* for the computer described in *item 1.*

a. $(+101110) + (-7)$ b. $(+110111)(-100)$

c. $(+111111)/(-7)$ d. $(+111101100) - (-62)$

computer circuits

Many people have a tendency to become overawed when they see a large-scale computer in action. If you happen to be one of these people, it should be comforting to know that the entire machine is composed of *many repetitions* of only a *few very simple circuits*. We are concerned with only two types of signals: *data* and *control*. There is some overlap, but as a rule, we have different circuits for handling the different signals.

5-1 DATA CIRCUITS

All data are in the form of binary *ones and zeros*, and these digits are represented, respectively, by the presence or absence of an electric pulse. Our data circuits are concerned with *moving* and *storing* the data pulses. In the process, care must be exercised to keep the data pulses in a *uniform* shape and amplitude. The primary data circuits then are flip flops, amplifiers, inverters, clampers, and limiters.

Clamper Circuits

The clamper is a circuit which establishes the starting *reference* for a signal. Such circuits are necessary because coupling circuits tend to shift the dc reference levels. Clamper circuits may be formed from either diodes or transistors. The clamper may be either positive or negative, and either type may be unbiased or biased to any desired potential. Some manufacturers refer to the clamper as a *dc restorer* (DCR) because of its function in restoring the desired reference level. Figure 5-1 illustrates both positive and negative unbiased clampers.

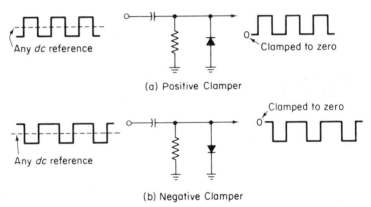

(a) Positive Clamper

(b) Negative Clamper

Fig. 5-1. Unbiased Clampers.

The diode in the *positive* clamper conducts when the input is *negative* with respect to ground. This charges the capacitor as indicated and lowers the reference level on the input signal. The output is *developed* across the resistor while the diode is *cut off*. This clamper establishes the *most negative position* of our positive going signal at a reference level of *zero*.

Our *negative* clamper has the diode reversed, which means that the diode *conducts* when the input is more *positive* than ground. This charges the capacitor as indicated and raises the reference level. Our output is developed across the resistor while the diode is cut off. This clamper establishes the *most positive portion* of our signal at a reference level of *zero*.

We may establish the reference (starting) level for our signal at *any* desired value by *biasing* these clampers. Figure 5-2 illustrates how this is accomplished.

Our capacitor now charges to the bias level plus the reference level of the input. The *output reference* then becomes the same as the *bias level* on the diode.

When discrete components are used, we generally find that diodes are used in the clamper circuits. This is an economy measure because diodes are

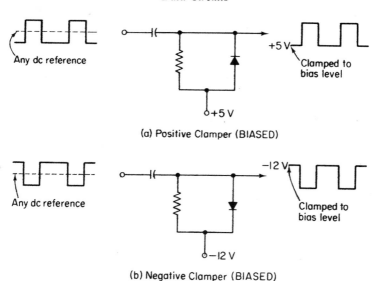

(a) Positive Clamper (BIASED)

(b) Negative Clamper (BIASED)

Fig. 5-2. Biased Clampers.

less expensive than transistors. When integrated circuits are used, we may find that most clampers use transistors instead of diodes.

Limiter Circuits

Limiters are necessary in computers in many places to *prevent* the maximum signal amplitude from exceeding the specified logic level. The designer decides what voltage level will be used to represent binary ones and what level will be used ro represent binary zeros. At various points in the circuit, such as the output of some amplifiers or flip flops, the logic level could well be exceeded by a significant amount. The designer inserts limiter circuits at these critical points. We have *positive limiters to limit the maximum swing* of positive signals and *negative limiters to limit the maximum swing* of negative signals. Figure 5-3 illustrates the operation of both positive and negative limiters.

In either case, the *maximum amplitude* of the output signal is limited to the *bias level*. Our outputs are developed across the diode while the diode is cut off. When the input amplitude reaches the bias level, the diode conducts and prevents any further excursion.

Flip Flop Circuits

Flip flop circuits are provided throughout our computer system to *temporarily hold* the binary bits as they are moved and manipulated. A flip flop

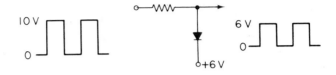

(a) Positive Limiter

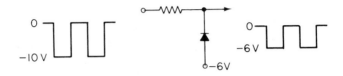

(b) Negative limiter

Fig. 5-3. Limiter Circuits.

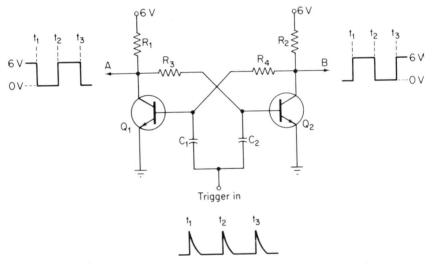

Fig. 5-4. Basic Flip Flop.

is essentially a *bistable* multivibrator with a few added features. A simple flip flop is shown in Figure 5-4.

In this circuit, *one* of our transistors is conducting at saturation, and *the other* is cut off. This is a *stable* condition, and it will hold until it is *upset* by an external signal. The wave shapes from the two collectors indicate that Q_1 was cut off and Q_2 conducting prior to t_1.

With Q_1 *cut off*, its collector voltage is 6 V because there is no current through R_1. This places forward bias on the base of Q_2, which keeps Q_2 saturated. The collector current from Q_2 causes the full 6 V to be dropped across R_2, and this holds the collector of Q_2 at 0 V. This same zero potential couples across R_4 to the base of Q_1, and holds it in a cutoff condition.

At t_1, we apply a positive trigger pulse through C_1 and C_2 to the base of *both* transistors. It has no effect on Q_2 because Q_2 is already saturated. The forward bias on Q_1 allows it to conduct with a resulting drop in collector potential. The decrease in Q_1 collector voltage couples across R_3 and reduces the forward bias on Q_2. The collector of Q_2 rises. This increase in potential couples across R_4 and increases the forward bias on Q_1.

In a very short time after we apply the trigger (t_1), we have Q_1 *conducting* at saturation and Q_2 *cut off*. This is another *stable* state that will hold until another *input* trigger is applied. While in this state $(t_1$ *to* $t_2)$ we have 0 V at the collector of Q_1 and 6 V at the collector of Q_2. This condition is completely *reversed* with the application of each input trigger pulse.

We generally designate one output of a flip flop as the "*zero*" output and the other as the "*one*" output. *In Figure 5-4*, if we consider output A as the zero output and output B as the one output, the flip flop is storing a one from t_1 to t_2. Output A is high from t_2 to t_3 which means that we stored a zero during this period.

The stable condition which produces a *high output from the one side* of the flip flop we call "*set*." Therefore, when we store a one, we *set* the flip flop. We refer to the opposite stable condition as either "*clear*" or "*reset*". When we store a *zero*, we clear the flip flop, and the output from the zero side becomes *high*.

The flip flop in Figure 5-4 has a *slow* reaction time. We can make it *faster* by two simple modifications. We can place *bypass capacitors* across R_3 and R_4 and use *diodes* to steer the trigger input to the *off transistor* only. While doing this, we might as well apply a small voltage to the base circuits and guarantee that only one transistor can conduct at any given time. Our improved flip flop is shown in Figure 5-5.

The negative 2 V at the junction of R_5, R_6, and R_7 holds a *reverse bias* on the *cut off* transistor until this bias is cancelled by an input positive trigger. This -2 V also aids the steering circuit which is composed of R_7, R_8, D_1, and D_2. A small current exists from the -2 V to ground, and this places a small negative potential on the anodes of the two diodes.

If we assume a condition which exists when the flip flop contains a one, Q_2 *is cut off* and Q_1 *is conducting*. The base of Q_2 is -2 V, and the collector is $+6$ V. The base of Q_1 is some positive value, and the collector is 0 V. D_1 is held cut off by 8 V of reverse bias, $+6$ V on the cathode and -2 V on the anode. The bias on D_2 is very near zero.

When we apply a *positive trigger* through C_3, the plates of both diodes be-

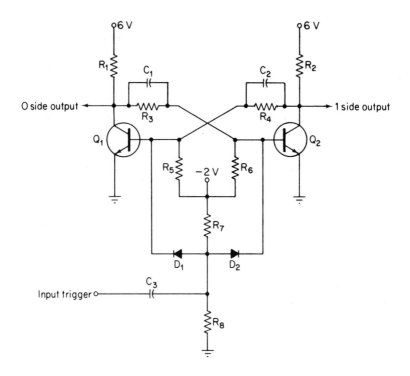

Fig. 5-5. A Faster Flip Flop.

come positive. D_1 remains *cut off* because its cathode is still more positive than the plate. D_2 will *conduct* as a result of the forward bias from the -2 V on the cathode and the positive pulse voltage on the plate. The resulting diode current is upward through R_6; and it develops a positive potential on the base of Q_2. This triggers Q_2 *into conducting* and its collector potential drops. The drop in potential couples through C_2 to the base of Q_1. Q_1 *cuts off*, and Q_2 goes to *saturation*.

This action has *reversed* the condition of our steering diodes. D_2 now has 8 V of reverse bias, and the bias on D_1 is near zero. The next positive trigger that is applied at the input will be directed to the *base of Q_1*. So, the steering diodes are *directing* the trigger pulses to the base of the *cut off* transistor.

A steering circuit for a negative trigger pulse is the same as the positive trigger steering circuit except for the fact that both *diodes are reversed*. This is necessary, of course, because the polarities of all voltages are reversed. This is illustrated in Figure 5-6.

When Q_1 *is conducting*, D_1 is blocked, and the negative trigger is directed to the *base of Q_2*. When Q_2 is conducting, D_2 is blocked, and the pulse is routed to Q_1.

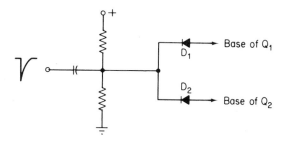

Fig. 5-6. Negative Pulse Steering Circuit.

Thus far, we have shown only one input to our flip flop, but we usually find *three inputs*. The *trigger* input, as we have observed, reverses the stable condition of the flip flop. One trigger pulse stores a one, and the next trigger pulse stores a zero. Doesn't it strike you that this would be an easy way to obtain the *ones complement* of a binary number? Well, it is, and that is how the computer accomplishes it. A register of six flip flops may contain the digits 101010. One single trigger pulse applied to all six flip flops at the same time will *reverse the condition* (complement) of each flip flop. The number is then 010101.

The other two inputs are the *set input* and the *clear input*. An activating signal at the *set* input stores a *one*. A *zero* is stored when an activating signal is applied to the *clear* input. The *trigger* input *complements* the binary bit in the flip flop. Notice that it is impossible to have a *neutral* flip flop. If it has voltage applied, it must contain either a zero or a one. Figure 5-7 shows the three flip flop inputs.

Regardless of the condition of the flip flop when the input arrives, a set pulse will store a one; the clear pulse will store a zero; and the trigger pulse will complement the number.

It is *important* that our signals have a uniform amplitude and a uniform reference level. For this reason we frequently find *limiters* on the outputs of flip flops, and sometimes, we find *both* limiters and clampers. Figure 5-8 illustrates the use of both limiters and clampers.

With this arrangement, we assure *both* a fixed reference level and a maximum amplitude. When the flip flop contains a one, Q_2 is cut off, and the collector voltage could be as much as $+20$ V. The 6 V bias on the cathode of D_2 will cause D_3 to conduct when its anode exceeds a positive 6 V. This *limits* the maximum amplitude to $+6$ V. D_4 has ground at its anode and conducts only when the cathode becomes more negative than ground. Thus, D_4 *establishes the reference level* of the 6 V output as 0 V. D_1 and D_2 accomplish the same thing with the zero side output; D_1 is the *limiter*, and D_2 is the *clamper*.

The flip flop is repeated thousands of times in our *data circuits*, but it is

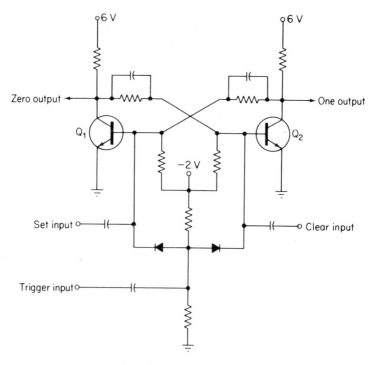

Fig. 5-7. Three Inputs to a Flip Flop.

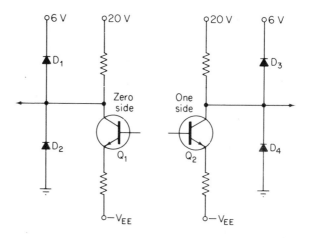

Fig. 5-8. Limiting and Clamping the Flip Flop Outputs.

also used extensively in *control circuits*. Among other things, it is used as a frequency divider in the timing circuits, and it is the primary building block in the counter circuits.

Flip Flop Latch Circuit

We use the flip flop latch extensively in computer circuits to do the jobs formerly reserved for flip flops. It resembles a flip flop in many ways, but the only circuit components are resistors, transistors, and diodes. Like a flip flop, it can be locked in either of two states in order to store a zero or a one. Figure 5-9 is one version of a flip flop latch.

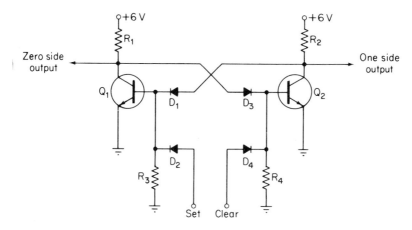

Fig. 5-9. Flip Flop Latch Circuit.

As the circuit is drawn, either transistor can be conducting and the other cut off. The quiescent state is of no importance in the basic operation of our circuit. Let's continue with our policy of representing a zero with 0 V and a one with +6 V.

When we wish to store a one, we apply a +6 V pulse to the set input. This causes D_2 to conduct, and places +6 V on the base of Q_1. Q_1 is forward biased, causing a strong collector current through R_1. Q_1 collector voltage drops to 0 V and cuts off D_3. The clear pulse is *not* present which is equivalent to 0 V, and this holds D_4 cut off. There is no current through R_2, and 0 V is on the base of Q_2. Q_2 is reverse biased and held in a cut off state. There is *no* current through R_2 which holds the collector voltage at +6 V.

The +6 V on the collector of Q_2 is our one side output. This +6 V also places forward bias on D_1 and causes it to conduct. The current through D_1 is also through R_3 from ground. This holds a positive potential on the base of Q_1 after the set input has returned to a 0 V level. We sometimes call this path the *latch back* because it locks the latch in the one state.

When we remove the one, we store a zero, and this is done by applying +6 V to the *clear* input. This causes D_4 to draw current through R_4 and places a positive potential on the base of Q_2. Q_2 conducts, and the collector voltage drops to 0 V. D_1 is now reverse biased, D_2 is already cut off (set input = 0 V), and this places 0 V on the base of Q_1. Q_1 cuts off, and its collector potential rises to +6 V.

The +6 V on the collector of Q_1 is our zero side output, and it also locks our latch in the zero state by holding forward bias on D_3. D_3 draws enough current through R_4 to hold Q_2 in conduction after the clear input has returned to a 0 V level.

The flip latch can be designed for either positive or negative logic, and it can be used to pass signal pulses as well as to store data. If we need a start pulse for initiating an action, a constant level to keep it going, and a stop pulse to shut it off, the start pulse is applied to the set side and the stop pulse to the clear side. The start pulse gives us +6 V at the one side output, and the stop pulse brings this level back to 0 V.

When we need to transfer a pulse or a level, this can be accomplished by applying a set and a clear at the same time. This action causes both outputs of the flip flop latch to assume the same voltage level for the duration of the inputs. With the circuit in Figure 5-9, a +6 V at both inputs provides a 0 V at both outputs. The circuit can be redesigned to give two high levels when two lows are applied to the inputs.

Amplifier and Inverter Circuits

In some computer systems, we find that there is only a *difference* of 0.3 V between a zero and a one. This makes the *amplifier* a very important and frequently used circuit. We encounter the *common emitter* configuration most often because it porvides amplification of voltage, current, and power. However, the common base and common collector configurations are used also. All *three* configurations are illustrated in Figure 5-10.

The three transistors shown in Figure 5-10 are NPN but the PNP may be used as well. The *only difference* between an NPN and a PNP amplifier is the polarity of the bias.

Figure 5-11 *summarizes* the basic characteristics of the *three* configurations.

Notice that while the common emitter provides gain of voltage, current, and power, it also has an output 180° *out of phase* with the input. When *one transistor* is used alone in this configuration, our amplifier becomes an *inverter*. There are many cases when inversion of the signal is *desirable*, and the common emitter serves this purpose very well.

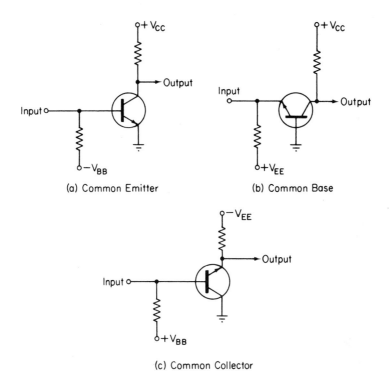

(a) Common Emitter (b) Common Base

(c) Common Collector

Fig. 5-10. Amplifier Configurations.

Characteristics	Common base	Common emitter	Common collector
Input resistance	$30 - 150 \ \Omega$	$500 - 1500 \ \Omega$	$20 - 500 \ k\Omega$
Output resistance	$300 - 500 \ k\Omega$	$30 - 50 \ k\Omega$	$50 - 1000 \ \Omega$
Voltage gain	$500 - 1500$	$300 - 1000$	Less than 1
Current gain	Less than 1	$25 - 50$	$25 - 50$
Power gain	$20 - 30$ db	$25 - 40$ db	$10 - 20$ db
Phase of output	Same as input	Inverted	Same as input

Fig. 5-11. Characteristics Summary.

When we do not wish inversion of the signal, we have *three* choices: common base, common collector, or *two* common emitters. Figure 5-12 shows the common emitter used as *both* an inverter and an amplifier.

When we use an *inverter*, our primary interest is a 180° *phase shift* of the signal from input to output. The resistor in series with the input *attenuates* the input to compensate for the *gain* of the transistor. This enables our output to be the *same* amplitude and 180° *out of phase* with the input.

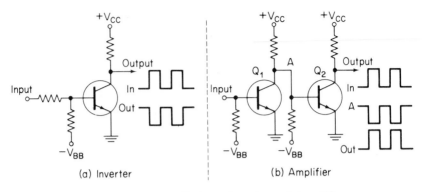

(a) Inverter   (b) Amplifier

Fig. 5-12. Common Emitter; Inverter and Amplifier.

In Figure 5-12B, we have *two* common emitters with the output of the *first* being the input to the *second*. The input signal is amplified and inverted through Q_1 so that the signal at *point A* is 180° out of phase with the input. The signal at point A is amplified and inverted again through Q_2. The overall result is a *double inversion* of the input signal, and the final output is *in phase* with the input.

Amplifiers are called by *several* names according to how they are used. The *final amplifier* in a unit generally feeds its output into a transmission line leading to another unit. The signal at this point must be provided with additional power to compensate for the losses in the line. We generally find that these amplifiers are called *buffers or drivers*.

Another amplifier is the *first stage* encountered by a signal as it enters a unit from a transmission line. We sometimes find these amplifiers designated as *receivers*. *Sensing amplifiers* are used to accept signals that are read out of memory storage.

5-2 CONTROL CIRCUITS

The control circuits are concerned with *synchronizing all actions*. This is done by initiating data movements, arithmetic operations, data manipulations, etc., in harmony with a *master timing system*. The principal circuits encountered in the control section are oscillators, flip flops, single shot multivibrators, delay lines, and gating devices.

The Clock System

In all computer systems we will find an *oscillator* which is the *source* of all timing signals. Practically every action in the system is initiated and controlled by pulses which originate in this oscillator. Since one oscillator can drive only a *limited* number of circuits, most systems contain *several* timing oscillators.

We call a timing oscillator a *clock*, and the pulses which originate here are *clock pulses*. Since all actions must be synchronized to one frequency, we have *one master clock* and *several slave clocks*. The master clock forces its frequency upon each of the slave clocks so that *all* clock pulses occur at the *same rate*.

Clock oscillators must have a highly *stable* operating frequency. We generally find that the master oscillators are *either* crystal controlled or tuning fork controlled oscillators. Figure 5-13 is a sample oscillator which could be used as a master clock.

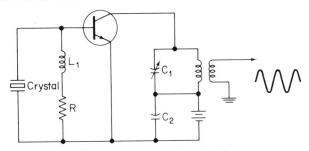

Fig. 5-13. Crystal Controlled Oscillator.

The output of the oscillator is a continuous repetition of a *sine wave* at a *fixed* frequency. We will *amplify* these sine waves, *square* them off on top, remove the negative alternations, and *differentiate* the remaining positive square waves. A limiter will now remove the nagative pulses, and this leaves us with a *pulse* for each *cycle* of the oscillator output. The wave shapes in Figure 5-14 illustrate the steps in these alternations.

Our final clock pulse (F) is shaped into a pulse of standard amplitude and width by amplifying and limiting the top.

These clock pulses are distributed throughout the system, and every computer action will be initiated by one of these pulses.

Assuming that our clock frequency (from the oscillator) is 3 MHz, the pulser recurrence time (PRT) or time between pulses is:

$$t = 1/f$$
$$t = 1/3 \times 10^6$$
$$= 0.333 \ \mu s$$
$$= 333 \ ns$$

The pulse recurrance frequency (PRF) is the same as the clock frequency, 3 MHz.

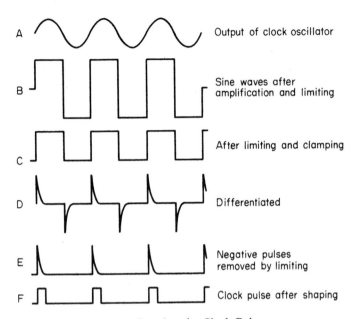

A — Output of clock oscillator

B — Sine waves after amplification and limiting

C — After limiting and clamping

D — Differentiated

E — Negative pulses removed by limiting

F — Clock pulse after shaping

Fig. 5-14. Forming the Clock Pulses.

This is our *master clock*, and we have too many circuits for one clock to handle. We can use free running *blocking oscillators* to assist in this task. The free running frequency of these blocking oscillators will be slightly *less* than our master clock frequency. When we *trigger* them with outputs from the master clock, each blocking oscillator is *synchronized* to oscillate at the same frequency as the master clock. Each blocking oscillator then becomes a *slave* clock, and each of them can control a specified number of circuits. In a computer, many things can happen in a 333 ns interval, and we must make provision for triggering actions that occur *between* clock pulses. We can do this with *delay lines* and single shot multivibrators.

Delay Lines

The artificial delay line is illustrated in Figure 5-15.

When we apply a pulse to the input of this line, it propagates down the line in a period of time determined by the square root of $L \times C$ in each section of line.

$$t = \sqrt{LC}$$

Where t is the time in *seconds* for a signal to travel through one section, L is the inductance of that section in *henries*, and C is the capacitance of that

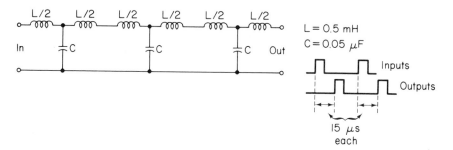

Fig. 5-15. A 15 μs Delay Line.

section in *farads*. The delay in *each section* of the line in Figure 5-15 is:

$$t = \sqrt{LC}$$
$$= \sqrt{0.5 \times 10^{-3} \times 0.05 \times 10^{-6}}$$
$$= \sqrt{25 \times 10^{-12}}$$
$$= 5 \times 10^{-6} = 5 \ \mu s$$

The line has *three* sections; therefore, the total delay is $3 \times 5 \ \mu$s or 15 μs. The line may be *tapped* at various places to obtain delays in smaller increments.

Single Shot Multivibrator

The single shot multivibrator has *one* stable state. An input *trigger* pulse will cause it to *switch* to the unstable state. We determine the *duration* of the unstable state by the *RC time constant* of the circuit. Consider the circuit in Figure 5-16.

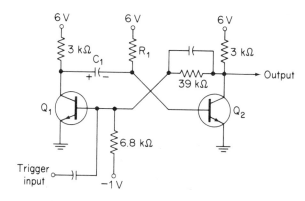

Fig. 5-16. Single Shot Delay Circuit.

The *stable state* is with Q_2 saturated and Q_1 cut off. C_1 is charged to 6 V in the direction indicated as a result of base current from Q_2. When we apply a positive *clock pulse* at the input, Q_1 will conduct. The drop in Q_1 collector voltage causes C_1 to start *discharging* through R_1. This places a negative on the base of Q_2 and drives it to cut off. This *unstable state* will remain in effect until the charge on C_1 diminishes to almost zero.

When the charge on C_1 approaches 0 V, Q_2 starts to conduct. The drop in collector potential cuts off Q_1. Q_2 goes into saturation, and the single shot is back in its *stable* state.

During the time that Q_2 is cut off, a *positive* square wave is taken from its collector, and a negative square wave is available at the collector of Q_1. Differentiation of the negative square wave produces a positive trigger pulse coincident with its *trailing edge*. This pulse is equivalent to a clock pulse that has been *delayed* for 250 ns. We can *change* the delay time by altering the value of *either* R_1 or C_1. If a manually *variable* delay is desired, we may use a variable capacitor.

Sometimes we find the single shot multivibrator used as a *data storage device*. If the storage is only needed for a very *brief* period of time, this is a very efficient way to do it. If we use the circuit in Figure 5-16 to *store* a binary one instead of delaying a clock pulse, the input *data* will be the trigger pulse. A binary *one* will cause *switching* to the unstable state; a binary *zero* causes *no* action. After the arrival of our one bit, Q_2 will be *cut off* for 250 ns. During this time, we can remove a binary one from the collector of Q_2. At the end of 250 ns, the circuit *reverts* to the stable state, which is equivalent to switching from a *one to a zero*. So, if we need to store information for a *brief* period of time, the single shot will do it with a minimum of circuits, and it automatically *clears itself* after holding the information for a specified time.

AND Circuit

The AND circuit is a gate with *two or more inputs*, and we must have *all inputs activated* in order to obtain an output.

We have several varieties of AND circuits. A simple version is illustrated in Figure 5-17.

In the quiescent state, *both* diodes are conducting in the direction indicated. Our total current is about 12 mA which gives us a full 6 V drop across R_1. Practically speaking, this leaves 0 V at the output, point C. Positive inputs are required at point A *and* B in order to cut off the diodes. If either diode is cut off, the other will carry more current. Therefore, even when only one diode conducts, our output remains as 0 V.

Applying a positive 6 V to *both* A and B at the same time will *stop* all diode current. With no current through R_1, point C is the same potential as the applied voltage which, in this case, is 6 V.

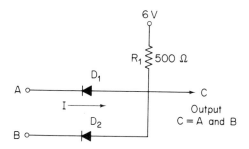

Fig. 5-17. Circuit of a Diode AND Gate.

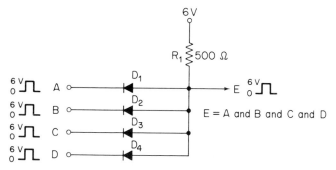

Fig. 5-18. Four Input AND Circuit.

We frequently refer to input lines as legs. When more inputs are added, we must have a positive 6 V for *each* leg in order to obtain our +6 V out. This is illustrated in Figure 5-18.

Either diode is capable of furnishing the 12 mA of current necessary to keep the output at 0 V. Therefore, we must have *four* separate inputs, all at the same time, in order to have a positive 6 V at point E.

If we allow 0 V to represent a binary zero and +6 V to represent a binary one, some *uses* for the AND gate become apparent. For instance, we can use a two legged AND gate to *move* a data bit from one storage location to another and have the movement take place at a *predetermined time*. This is shown in Figure 5-19.

Leg A is connected to a data input line. The incoming string of binary bits have spacing *intervals* equivalent to the PRT of the clock. When a one bit *coincides* with a clock pulse, both diodes cut off, and we have 6 V at point C. This effectively *transfers the one* bit to the output. When a binary zero occurs, the clock pulse cuts off D_2, but D_1 continues to conduct, and holds point C at zero volts. This effectively *transfers the zero* bit to the output.

Data are represented by voltage *levels* as well as pulses. When a binary bit is stored, it establishes a level which holds as long as the bit is there. If we

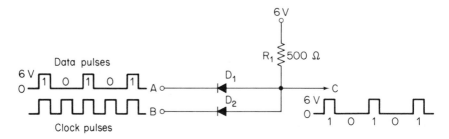

Fig. 5-19. Gating Data With an AND Circuit.

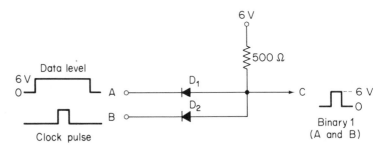

Fig. 5-20. AND Gating a Data Level.

use this level on one leg and a clock pulse on the other, we can still move the binary bit through our AND gate. This is illustrated in Figure 5-20.

During the time that the binary one is stored in our previous stage, a $+6$ V *level* is maintained on leg A. We call this a *conditioning level*. Leg A is conditioned, but the output remains at 0 V until a positive 6 V is applied to leg B. During the *clock pulse*, D_2 is cut off, and the output rises to 6 V. The level is still present on leg A, but when the clock pulse *terminates*, the output returns to zero. During the clock pulse, we *pass* the binary one through our AND gate.

When the stored bit is a zero, the data level is 0 V. The 0 V on leg A *allows* conduction, and the output is held at a zero level.

AND circuits are frequently composed of *transistors* as well as diodes. We also find *hybrid* arrangements containing both diodes and transistors. Figure 5-21 illustrates a hybrid AND gate.

This common collector (emitter follower) configuration produces 0 V at point D in the *static* condition. *Each* of our input diodes is forward biased and conducting. This keeps the base of the transistor at approximately the *same* dc level as the emitter (0V). In order to obtain a *positive output* at point D, we must *apply* a positive to A, B, and C and cut off *all three* diodes. About 6 V is required to cut off each diode. When we apply a $+6$ V pulse to A, B, and C *simultaneously*, all diodes cut off, the transistor is forward biased, emitter current increases, and point D rises to $+6$ V.

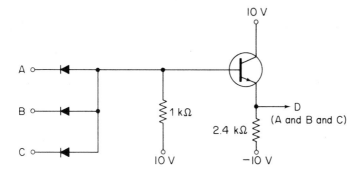

Fig. 5-21. Combined (Diodes and Transistor) AND Gate.

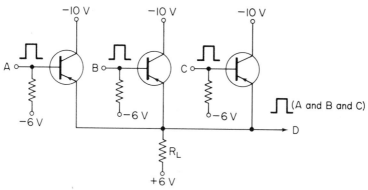

Fig. 5-22. Transistor AND Gate.

Figure 5-22 shows an AND gate that uses only transistors.

In the static condition, we have *all transistors conducting*, and the output at point D is zero. Since the load resistor is common to all transistors, the output will remain at 0 V as long as *any one* of them conducts. A positive pulse of 6 V is required to cut off *either* of the transistors. When we apply +6 V to A, B, and C simultaneously, all transistors are off, current through R_L ceases, and we have +6 V at point *D*.

The AND gate performs a *logical function* by making a decision. The designer has determined that a certain job should be initiated *each time* the specified conditions are met. *Each* of these conditions raise the level on one leg of the AND gate. When *all conditions* are met, the AND gate provides an output to start the action. Here are some examples of logical AND functions.

1. If you *attend 90%* of your classes AND maintain a *C average* you will *graduate*.

2. If you *attend all* classes AND maintain a *B+ average* you will make the *honor roll*.

OR Circuit

The OR circuit is a gate with *two or more inputs,* and if *one or more* inputs are activated we have an output. A basic OR circuit is illustrated in Figure 5-23.

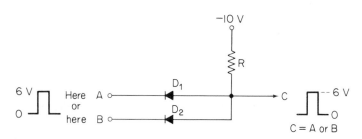

Fig. 5-23. Diode OR Circuit.

When we have no input, *both* diodes are *conducting* at a level which holds a *zero* potential at point C. When we apply a positive pulse to *either* (or both) A or B, there will be an increase of current through R, and the output rises to the same positive value as the *most positive* input.

An OR circuit may have *several* inputs, and when we apply a positive pulse to *any leg* (or in any combination), the circuit produces a *positive* pulse in the output. Figure 5-24 illustrates a combination diode-transistor OR gate.

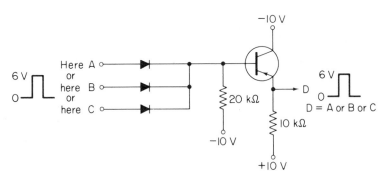

Fig. 5-24. Combination (diode-transistor) OR Gate.

When we have 0 V on *all three* legs, the transistor and all diodes are conducting and the emitter current is holding the output at 0 V. Applying a *positive* pulse to either A, B, or C will cause that diode to increase conduction. The resulting rise in positive potential at the base of the transistor causes it to conduct *less.* The *decrease* in emitter current drops less voltage across R_L, and the output rises to the input level.

Figure 5-25 illustrates an OR gate formed by three transistors.

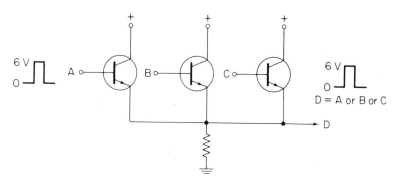

Fig. 5-25. Transistor OR Gate.

With *no* inputs, *all* transistors are *cut off*, there is no current through R_L, and our output is at a *zero level*. When we apply a positive to *either* A, B, or C, we will cause the associated transistor to conduct. The resulting current causes a *voltage drop* across R_L, and the output positive rises to approximately the same amplitude as the input.

The OR circuit also performs a *logical* function; it makes a decision. It provides an *output* each time *any one* of a number of specified conditions occurs. Our output will initiate some job which should be done each time the conditions are right. Here are some examples of logical OR functions:

1. If I can *raise taxi fare* OR *get the car keys*, I'll attend the opening.

2. If you *call*, *write*, OR *wire*, your reservation is *guaranteed*.

It is not necessary that you call, write and wire; *any one* of these actions will suffice. However, if *you* write, and *your secretary* calls, and *your mother* sends a telegram, no harm is done. The reservation is still guaranteed.

NAND Circuit

The NAND gate is another *logical* function. With this circuit we provide an output at all times *except* when specified restraining conditions exist. For example:

1. You will be in class at *all times* EXCEPT when you are *sick and have a doctor's certificate*.

2. Dinner is served at *eight o'clock* sharp EXCEPT on the *cook's day off* and when *the wife is out of town*.

The NAND circuit is illustrated in Figure 5-26.

When *either* A or B has a low input, *neither* transistor can conduct. Under these conditions, we have a constant 6 V level at point C. The *only* way we can *stop* this positive output is to apply a positive pulse to *both A* and *B*

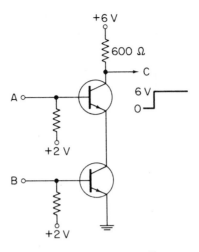

Fig. 5-26. Transistor NAND Gate.

simultaneously. When *both* A and B are positive, *both transistors* conduct, and the level at point C drops to 0 V. So, we have a *high* output at all times *except* when *both* A and B are high.

NOR Circuit

The NOR circuit is still another *logical* circuit. It provides an output *only* when both of our inputs are *low*. Examples of logical NOR functions are:

1. When *neither bus* NOR *subway* is convenient, I use *my own car.*

2. When there is neither *wind* NOR *surf*, I *enjoy my canoe.*

So, the NOR circuit *permits* us to perform an action when, and only when, *none* of the specified conditions are present. The NOR circuit is illustrated in Figure 5-27.

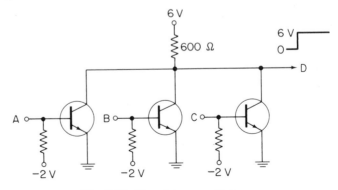

Fig. 5-27. Transistor NOR Gate.

When we have *no* input (A, B, and C all low), the transistors are *all* biased below *cut off,* and the output is a constant 6 V level. We may think of this as the *permissive* condition. When we apply a positive to either A, B, or C (or any combination thereof) the selected transistor (or transistors) will conduct. The resulting collector current will drop the 6 V across the resistor and the output becomes 0 V.

Exclusive OR Circuit

The exclusive OR (XOR) logic circuit has *two* inputs and *one* output. It will pass a signal when *one* input is *high* and the *other* is *low.* We have an example of this circuit in Figure 5-28.

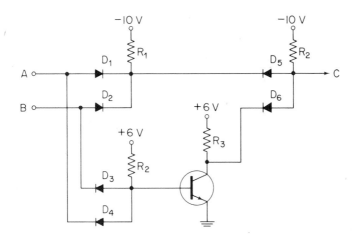

Fig. 5-28. Exclusive OR Gate.

Actually the XOR is a *comparison* function; we get an output *only* when A is *different* from B. We may also think of it as a *half adder.* It will take the *sum* of the bits at points A and B without considering the *carry.*

D_1, D_2, and R_1 form an OR circuit. An AND circuit is formed by D_3, D_4, and R_2. The transistor is an *inverter,* and D_5, D_6, and R_4 compose another AND circuit.

Consider the circuit when *both A* and *B* are 0 V. D_1 and D_2 are cut off, and this allows D_5 to conduct. D_3 and D_4 are *both* conducting which places 0 V on the base of Q_1. Q_1 is *cut off* and its collector potential is $+6$ V which holds D_6 cut off. The current from D_5 drops 6 V across R_4 and leaves 0 V output at point C.

When *both* inputs are $+6$ V, D_1 and D_2 *conduct,* placing $+6$ V on the cathode of D_5 and *cutting it off.* Both D_3 and D_4 are cut off, and the potential on the base of Q_1 is $+6$ V. Q_1 *conducts* and the 0 V at its collector causes D_6 to conduct. The current from D_6 drops 6 V across R_4, and the output is *still* 0 V. So, as long as the two inputs are the *same*, our output is 0 V.

When A is $+6$ V and B is 0 V, D_1 is cut off and D_2 conducts. This causes D_5 to be cut off. D_3 conducts, and D_4 is cut off. This results in 0 V on the base of Q_1. Q_1 *is cut off*, and the collector potential is $+6$ V. This 6 V cuts off D_6. We have *no* current through R_4, and the output at point C is $+6$ V.

Reversing the order of input should produce the same result; 0 V at A and $+6$ V at B should give $+6$ V out. D_1 is cut off, and D_2 conducts. The resulting 6 V keeps D_5 cut off. D_3 is cut off, and D_4 conducts. The resulting 0 V causes Q_1 to *stay* cut off, and its collector voltage is $+6$ V. The $+6$ V holds D_6 cut off. We have *no* current through R_4 and the output at point C is $+6$ V.

Schmitt Trigger

The Schmitt trigger is frequently encountered in computer circuits. It is a circuit which gives us a *standard output level* as long as it has an activating input. A sample circuit is shown in Figure 5-29.

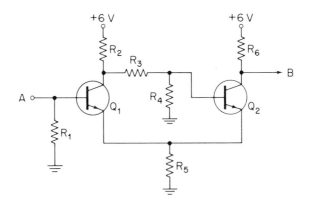

Fig. 5-29. Schmitt Trigger Circuit.

The principal use of this circuit is *wave shaping*. We can trigger it by applying almost *any type* of positive signal to point A. Our output at point B will be $+6$ V *as long* as the input signal is strong enough to keep Q_1 conducting.

The *quiescent state* of the Schmitt trigger is Q_1 *cut off and* Q_2 *conducting*. The output during the quiescent state is 0 V.

A negative input at point A has no effect. A *positive input* causes Q_1 to conduct, and the collector potential drops to 0 V. The 0 V couples across

R_3 and causes Q_2 *to cut off.* The collector voltage of Q_2 is 6 V, and this is our output at point B. The output will *retain* this 6 V level as long as the input keeps Q_1 *in the conduction state.* When Q_1 cuts off, Q_2 again conducts and reduces the output to 0 V.

5-3 TYPES OF LOGIC CIRCUITS

As the computer industry moves more into the realm of *integration,* more and bigger sections of circuits will be available in *standard packages.* The *type* of logic circuits selected by a circuit designer will be governed by the desired levels of 0 and 1, the standard of performance, and the cost per circuit.

RTL (Resistor-Transistor-Logic) Circuits

RTL circuits are manufacturered by *most major* integrated circuit manufacturers. They are quite simple and relatively inexpensive. The circuits are designed so that the binary zero is represented by *any* voltage level less than $+0.5$ V. The binary one is represented by voltages in *excess* of $+0.8$ V. The 0.3 V *difference* between a zero and a one is intended to prevent noise interference. Figure 5-30 is a sample inverter circuit from the RTL line.

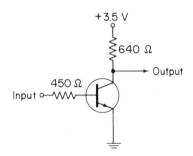

Fig. 5-30. Inverter Using RTL.

The *standard* circuit in the RTL line is the NOR circuit, but inverters and flip flops are also available. Regardless of the circuit, the components are *only* resistors and transistors.

DTL (Diode-Transistor-Logic) Circuits

The DTL is another popular line of integrated circuits. We find that these circuits are formed of *combinations* of diodes and transistors. The *standard* gate in this system is the NAND gate, but they can be combined to form ORs, ANDs, amplifiers, inverters, and flip flops. *Advantages* of the DTL circuits are

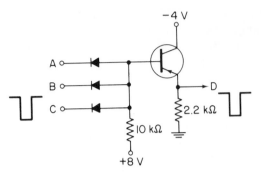

Fig. 5-31. OR Gate in DTL.

higher voltage levels, more amplification, and greater *differences* between the levels that represent zeros and ones. Figure 5-31 is a sample OR circuit in DTL.

TTL (Transistor-Transistor-Logic) Circuits

TTL circuits are more *expensive* than the lines mentioned previously. We can obtain them with *either* discrete components or in integrated packages. The higher cost is offset by the *higher speeds* offered by these circuits. The TTL circuits also provide strong drive power and a high immunity to noise interference. A sample integrated NAND gate is illustrated in Figure 5-32.

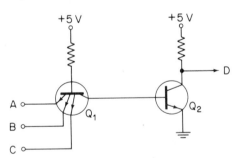

Fig. 5-32. TTL NAND Gate.

When *all* inputs to the *multiple* emitter transistor are high, Q_2 is saturated, and our output is 0 V. When we have a *positive* at A, B, or C (or any combination), Q_1 is saturated, and Q_2 is cut off. So, *any* positive in gives us a positive out.

Review Exercises

1. Name the two principal types of signals which circulate in a computer system.

2. How are binary digits represented in computer circuits?

3. Starting from left to right, draw a wave train to represent this binary number: 101011. Label the ones and zeros.

4. Under what conditions are binary bits represented by steady dc levels? Cite an example.

5. What is the relationship of a clamper and a DCR?

6. Draw a clamper circuit that will establish a −10 V reference for a +6 V pulse. Show input and output wave shapes with proper levels.

7. (a) Name the circuit in Figure 5-33.
   (b) Draw the output wave shape and indicate the proper levels.

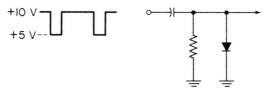

Fig. 5-33. Name the Circuit.

8. The output of the circuit in Figure 5-34 normally swings from +3 V to +20 V. Add two diodes to the circuit in a manner that establishes the output levels as 0 V and +6 V.

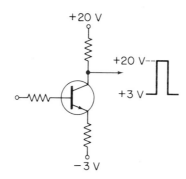

Fig. 5-34. Establish Output Levels.

9. (a) Name the circuit in Figure 5-35.
   (b) Draw the output and show the voltage levels.

10. What is the primary use of a flip flop circuit?

11. Name the three inputs to a flip flop.

12. Which input to a flip flop is used for:
    (a) Storing a one?
    (b) Removing a one?

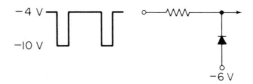

Fig. 5-35. What Type of Circuit?

(c) Complementing from one to zero?
(d) Complementing from zero to one?
(e) Storing a zero?

13. Under conditions where 0 V represents a zero and $+6$ V represents a one, we have a one stored from t_1 to t_2 and a zero from t_2 to t_3. Draw the outputs of the flip flop from t_1 to t_3 and label them according to the one and zero side output. Show the levels.

14. A flip flop is complemented by a series of 4 MHz clock pulses.
    (a) What is the frequency of the outputs?
    (b) What is the time interval between leading edges of the positive outputs?

15. (a) What job is performed by steering diodes?
    (b) How does this affect the switching speed of the flip flop?

16. Which basic amplifier configuration provides gain of voltage, current, and power?

17. Which configuration has a:
    (a) Current gain less than unity?
    (b) Voltage gain less than unity?
    (c) 180° phase shift?

18. What is the primary purpose of an inverter?

19. What is the difference between buffer, driver, and receiver amplifiers?

20. What is the purpose of the control circuits?

21. Describe a master clock oscillator.

22. Explain why most systems must have more than one clock.

23. Explain the difference between the free run frequency of master and slave clocks. Why is this necessary?

24. With a 1:1 ratio between oscillator sine waves and clock pulses, what frequency will produce clock pulse intervals of 250 ns?

25. A certain system has a clock PRT of 237 ns. The transmission lines between cabinets have 0.5 mH of inductance and 0.5 pF of capacitance per one foot section. What length line will provide a delay equal to the PRT of the clock?

26. What is the frequency of the clock in item 25?

27. The single shot multivibrator in Figure 5-36 is triggered by positive pulses. What components determine the duration of the unstable state?

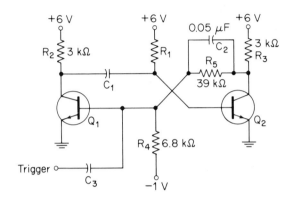

Fig. 5-36. Single Shot Multivibrator.

28. Why do we say an AND gate performs a logical function?

29. Draw a diode AND circuit which will provide +5 V out when it has +5 V on both input legs.

30. A four legged positive AND circuit is formed of a resistor and four diodes. We have only three inputs. What voltage level must we place on the fourth leg to enable the circuit to function as a three input gate?

31. Consider the circuit in Figure 5-37.
    (a) What type of circuit is it?
    (b) What type and how many pulses are required to activate it?
    (c) What is the output when activated?
    (d) What is the output when deactivated?

32. Draw an RTL NAND gate using two NPN transistors. Show activating inputs and resulting outputs.

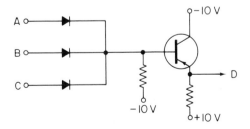

Fig. 5-37. Determine Type and Function.

33. In a system where $+6$ V represents a one and 0 V represents a zero, what are the conditions which produce a positive output from a NOR gate?

34. Describe the function of the XOR gate.

35. What is the principal use of the Schmitt trigger? How does it function?

logic
symbology

It is customary in the computer industry to represent computer *functions* by block diagrams. We call these functional diagrams "*logic diagrams*," and the symbols are "*logic symbols.*" The purpose of this chapter is to present the symbols and their functions. Since a single standard has *not* been agreed upon, we will consider two separate systems of symbols: the *uniform shape* and the *distinctive shape*. Neither of these may be considered standard, yet both are widely used. The industry is about equally divided as to which is most preferable. Only a few manufacturers adhere 100% to either system, but all are close enough that when we know both of these systems, we can read logic diagrams from nearly all manufacturers.

6-1 DEFINITIONS

A few definitions are essential to establish a common ground for this discussion.

LOGIC FUNCTION: An action such as storage, delay, comparison, etc., expressing a relationship between signal inputs and outputs.

LOGIC SYMBOL: A graphic representation of a *logic function*. (Notice that we specify function instead of circuit. In most cases, we do have a one to one relationship between logic symbols and logic circuits, but as long as the function is present, the existence or nonexistence of the actual circuit is unimportant. This distinction will become more and more important as computers become more integrated.)

LOGIC DIAGRAM: A diagram which uses logic symbols and notations to depict the details of signal flow and control in a system of two state devices. It is *not* necessary to show point to point wiring.

DISTINCTIVE SHAPE LOGIC DIAGRAM: A logic diagram where *common* functions are represented by enclosures of distinctive shapes and *less common* functions are represented by rectangles with internal labels denoting the function.

UNIFORM SHAPE LOGIC DIAGRAM: A logic diagram where *all* functions are represented by rectangles with internal labels denoting the function.

STATES IN BINARY LOGIC: The two physical states on each signal line are referred to as the "*0*" *state* and the "*1*" *state*. The zero state is the reference state or *inactive* state. The one state is the significant state or *active* state.

TAGGING LINES: Additional information contained within a logic symbol, such as unique identification of the symbol and means of locating the hardware.

6-2 ASSIGNING LOGIC LEVELS

Types of Logic

A system of logic where the *most positive* of two levels consistently represents a *one* we call *positive logic*. The less positive level then represents our zero. A system of logic which reverses this procedure we call *negative logic*. In negative logic the *most positive* of two levels consistently represents our *zero*, and the *less positive* represents the *one*. We frequently encounter a mixture of positive and negative logic within a system, and occasionally on the same sheet of logic. When both positive and negative logic are used on the same diagram, we call it *hybrid logic*.

Dual Functions

The function performed by a particular circuit is frequently determined by the *type* of logic. Suppose that we have a circuit that is capable of assuming only a $+2$ V and a -3 V level. Each input can be *either* $+2$ V or -3 V,

INPUTS		OUTPUTS
A	B	C
−3 V	−3 V	−3 V
−3 V	+2 V	−3 V
+2 V	−3 V	−3 V
+2 V	+2 V	+2 V

Fig. 6-1. Table of Combinations.

and the output can be either a +2 V or a −3 V. It is a two legged circuit, and the circuit functions according to the table of combinations in Figure 6-1.

If this circuit is in a system of *positive* logic, the −3 V is our *zero*, and the +2 V is our one. When we substitute these logic values for the voltage levels, our table of combinations becomes the *truth* table for an AND function. This is shown in Figure 6-2.

If we are dealing with *negative* logic, the −3 V is our *one*, and the +2 V is our zero. When we substitute these values, we obtain a truth table for an OR function as shown in Figure 6-3.

INPUTS		OUTPUTS
A	B	C
0	0	0
0	1	0
1	0	0
1	1	1

INPUTS		OUTPUTS
A	B	C
1	1	1
1	0	1
0	1	1
0	0	0

Fig. 6-2. Truth Table for a Positive AND Gate.

Fig. 6-3. Truth Table for a Negative OR Gate.

Therefore, any physical device which can be characterized by a table of combinations will have its function determined by the type of logic it appears in. A positive AND is a negative OR; a positive NOR is a negative NAND.

We have been dealing with logic levels which apply to positive logic by representing zeros with 0 V and ones with +6 V. We will maintain this convention as long as possible to keep confusion to a minimum. You will have fair warning before we change. Therefore, until further notice our:

1. Logic is positive.

2. Ones are +6 V.

3. Zeros are 0 V.

4. L(low) may be substituted for a 0.

5. H(high) may be substituted for a 1.

6-3 INDIVIDUAL SYMBOLS

We must learn to recognize about 15 individual symbols in *both* uniform and distinctive shape. In the following illustrations, the uniform shape will appear on your *left*, the distinctive shape on your *right*, and the truth table will represent *positive* logic. When only one symbol appears, it means that there is no distinctive shape for this symbol.

AND Function

The output of an AND assumes the one state when, and only when, *all* inputs are in the *one* state. See Figure 6-4.

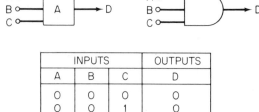

INPUTS			OUTPUTS
A	B	C	D
0	0	0	0
0	0	1	0
0	1	0	0
0	1	1	0
1	0	0	0
1	0	1	0
1	1	0	0
1	1	1	1

Fig. 6-4. The AND Function.

In this figure we have three legged AND gates, but any reasonable number of inputs can be used. In this case, D = A·B·C which is saying that all inputs must be a one in order to get a one out. Check it on the truth table. (The symbol, ·, is read as AND.)

OR Function

The output of an OR assumes the one state when *one or more* inputs are in the *one* state. The OR function is illustrated in Figure 6-5.

D = A + B + C. The sign, +, is read as OR, so D = A or B or C. The truth table shows that *any* combination of inputs that contains a one will produce a one in the output.

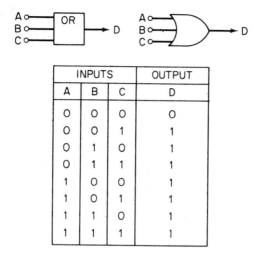

Fig. 6-5. The OR Function.

Exclusive OR (XOR) Function

This is a two legged input gate, and it provides a one out when the two inputs are *different*. We also call it a half add function and a compare function. The XOR function is illustrated in Figure 6-6.

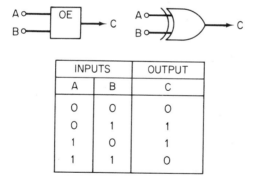

Fig. 6-6. The XOR Function.

The truth table shows that two 0 inputs, or two 1 inputs, will give us a 0 output. When one input is 1 and the other is 0, our output is 1. The formula is: $C = A\bar{B} + \bar{A}B$. The vinculum is read as NOT; $C = A$ and not B, or not A and B. In positive logic A is the *activating* signal (1) and NOT A ($\bar{A}$) is the *inactive* state (0).

Logic Negation or State Indicator

A *bubble* at the beginning or end of a signal line, similar to the small 0s on this page, is a symbol for logic *negation*. This symbol is never used alone but is combined with other symbols in the manner shown in Figure 6-7.

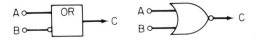

Fig. 6-7. Use of Logic Negation Symbol.

In the uniform shape OR, we have the bubble on leg B. B must be a *zero* for this leg to activate the function. Therefore, the formula is $C = A + \bar{B}$. The formula for the distinctive shape OR is $\bar{C} = A + B$. In the latter case, we have negation on the output which indicates that the activated output is a *zero*.

Electric Inverter

The electric inverter is a *single* input and single output function. When we put in a one, our output is a zero. When we put in a zero, our output is a one. This is a simple *inversion* of the electric signal. Figure 6-8 illustrates the inverter function.

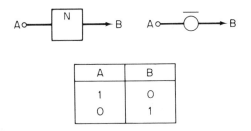

A	B
1	O
O	1

Fig. 6-8. Electric Inverter.

Inverters are widely used at points where logic changes from positive to negative or vice versa. In these cases, our inverted one changes its voltage level, but it is *still a one*. For example: We are using positive logic with O V being zero and $+6$ V being one. When we come to a point where we change to negative logic, we invert the signals. The $+6$ V (1) becomes O V (1), and the O V(0) becomes $+6$ V (0). Remember that in positive logic the one is the *relative high* of two voltage levels, and the zero is the *relative low* of the two voltage levels. The negative logic is exactly *opposite*. We have other level indicators to help us through the confusion of changing logic types.

Triangular Flags

A small *right triangle* (flag) may be used in the same manner as the bubble state indicator. The point of the triangle indicates the direction of our signal. If the flag is *solid* (black), the active level (1) is the relative *high*. Figure 6-9 combines the flags with inverter symbols.

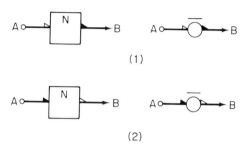

(1)

(2)

Fig. 6-9. Flags with Inverters.

In the first case, we are going *from negative* logic through the inverter to *positive* logic. A low (1) input gives us a high (1) output. A high (0) input gives us a low (0) output.

In example (2), we are going *from positive* logic through an inverter *to negative* logic. In this case, when we have a low (0) input, we have a high (0) output. When we have a high (1) input, we have a low (1) output.

These flag level indicators can be used on any or all signal lines at any point where there is another logic symbol. But like the bubble, they are never used alone. We will have other examples later, but for now let's go back to our positive logic and examine some more symbols.

NAND Function

The NAND function is a *combination* of the AND and the inverter. The symbol is the AND symbol with the logic negation bubble on the output. This is illustrated in Figure 6-10.

Our NAND function provides an *inactive* output of a *high* (*1*). The activating conditions are all inputs high (1) at the same time. The *activated* output is a *low* (*0*). The truth table shows that any combination of inputs that include a low (0) will give us a high (1) output.

NOR Function

The NOR function *combines* the OR and the inverter functions. The symbol combines the OR symbol with the logic negation bubble on the output. This is illustrated in Figure 6-11.

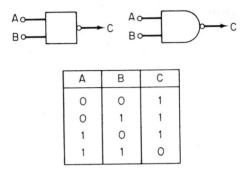

Fig. 6-10. NAND Function.

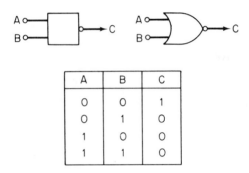

Fig. 6-11. NOR Function.

Our NOR function provides an *inactive* output of a *high* (*1*). It is activated by applying a high (1) to any input. Our *activated* output is a *low* (*0*). The truth table shows that we have a low (0) output at all times except when both inputs are low (0).

Flip Flop

The flip flop is a *storage* device which can temporarily hold a binary bit of information. It has a set, a clear (or reset), and generally a trigger input. It has a zero side output and a one side output.

When we apply a high (1) to the set input, we store a one in the flip flop. As long as the flip flop is *set*, we can take a high (1) from the *one side* and a low (0) from the *zero side*.

When we apply a high (1) to the *clear* side, we *store* a zero in the flip flop. As long as it is clear, we can take a high (1) from the *zero side* and a low (0) from the *one side*. The symbol is shown in Figure 6-12.

We sometimes find this referred to as a *complementing* flip flop. In this case, the logic probably uses other flip flops which have only set and clear

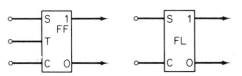

Fig. 6-12. Flip Flop Symbol.   **Fig. 6-13.** Symbol for a Flip Flop Latch.

inputs. At any rate, this flip flop has a complementing capability. When we apply a high (1) to the *trigger* input, the flip flop will *reverse* states. A complemented 0 is a 1, and a complemented 1 is a 0. We sometimes find the trigger input referred to as a toggle or a complement input, Applying a high (1) to *both* set and clear at the same time is another way to complement the flip flop.

Flip Flop Latch

The flip flop latch *resembles* a flip flop in function. It has two inputs which we can call *set and clear* or start and stop. It also has two outputs designated as *zero side and one side* output. It has *no* trigger input and *does not* complement. When we apply a high (1) to *both* set and clear at the same time, we will have a low (0) from both outputs for the duration of the inputs. The symbol for a flip flop latch is shown in Figure 6-13.

We *store a one* in the flip flop latch by applying a high (1) to the *set* input. While the one is there, we can take a high (1) from the *one side* output and a low (0) from the *zero side* output.

To *store a zero*, we apply a high (1) to the *clear* input. While the zero is there, we can take a high (1) from the *zero side* output and a low (0) from the *one side* output.

Single Shot

The single shot (SS) is the short name for our unistable multivibrator. It is also called a one shot multivibrator, or monostable multivibrator. The symbol is shown in Figure 6-14.

The principal use of this circuit is to provide a *delay* between the input signal and the trailing edge of the output. It is customary to show a square wave either inside or in the vicinity of the symbol. The square wave shows the *polarity* of the output and the *time* duration. In this case, we *trigger* the SS with a high (1), and the output is high for 0.5 μs. This is the duration of the delay or the length of time that a bit can be stored in the single shot.

Schmitt Trigger

The symbol for the Schmitt trigger is illustrated in Figure 6-15.

The Schmitt trigger (ST) is activated when the input high (1) rises *above*

Fig. 6-14. Single Shot   **Fig. 6-15.** Schmitt
Symbol.   Trigger Symbol.

a certain threshold (turn on) level. It will stay activated until the input drops *below* another specified threshold (turn off) level. The turn off level is lower than the turn on level. As long as the ST is on, the output is a high (1).

Amplifier

Figure 6-16 illustrates the symbols and truth table for our amplifier.

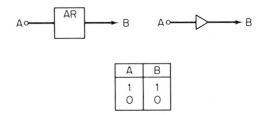

A	B
1	1
0	0

Fig. 6-16. Amplifier Symbols.

The amplifier increases the amplitude of either voltage, current, or power, and sometimes provides gain of all three. Notice in the truth table that we have *no* signal inversion. If we have a high (1) at the input, we have a high (1) at the output.

At one time, the distinctive shape for an inverter was the triangle with a state indicator on the output. This is shown in Figure 6-17A. The present distinctive shape symbol is that in Figure 6-17B, as we saw earlier.

A   B

Fig. 6-17. Former and Present Inverter Symbols.

Time Delay

Figure 6-18 shows symbols for time delay (delay lines) of *single* output and *multiple* outputs in both uniform and distinctive shapes. The indicated times show the amount of *delay* between the input and that particular output. In any case, a high (1) input gives us a high (1) output after the indicated time interval.

Example 1 shows a delay line with a single output, and the time *between*

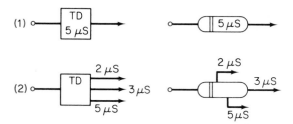

Fig. 6-18. Time Delay.

the leading edges of input and output is 5 μs. In the second example, we have three outputs. The time delay between the leading edge of the input and output varies according to which output line we use. The time, in each case, is shown near the output line.

Other Functions

There must always be provisions for showing functions that have *no* standard symbol. We cover this with the *general* symbol shown in Figure 6-19.

A rectangle of this type is used to represent any function that has *no* logic symbol of its own. This symbol will have a label to clearly identify the function it represents.

Buildup Symbol

This is a variation of our general symbol, and it is used to symbolize *redundance* of a particular function. For instance, an 8 bit computer word requires a temporary storage space. This storage area is a register composed of 8 flip flops which all function in the same manner. Symbolically we represent the register as shown in Figure 6-20.

The first section of the function is shown in detail, and each of the other sections are understood to be *identical* to the first. In this case, we show 8

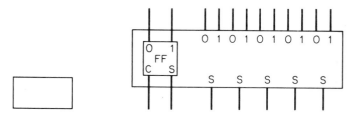

Fig. 6-19. General Symbol.

Fig. 6-20. Buildup Symbol.

flip flops with all inputs and outputs. The clear input is *common* to all 8 flip flops. We find variations of this symbol widely used for registers, counters, and various types of matrices.

Multiple Inputs

The multiple input provision is simply an extension of the input side of the logic symbol concerned. The side is *extended* as far as necessary to make room for all the inputs. Figure 6-21 illustrates multiple inputs for both AND and OR gates.

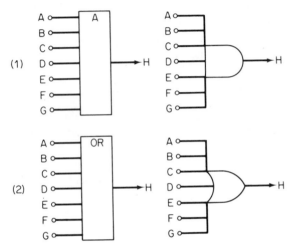

Fig. 6-21. Multiple Inputs.

Connector Functions

We frequently find circuit *connections* which combine the outputs of two or more circuits, and in doing so, they create *another* logic function. The function so created is either an AND or an OR. Figure 6-22 illustrates the symbology for connector functions.

In uniform shape logic, the connector function may be represented by placing *two* function designators *inside* the associated logic symbol as we see in the drawing. This means that a normal function is *followed* by a connector function. When necessary for clarity still another designator (A or OR) will be placed near the function.

In the distinctive shape logic, we find the connector function symbolized by a *small* logic symbol which *encloses* the junction.

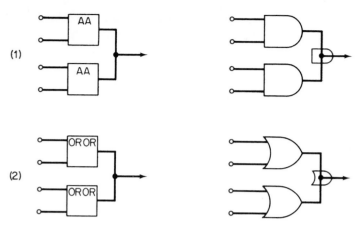

Fig. 6-22. Symbolizing Connector Functions.

Combined Functions

When separate functions are performed by the *same* hardware, we sometimes find the symbols *tangent* to one another as shown in Figure 6-23. Example 1 shows that *any* one of three inputs can *set* the flip flop. The

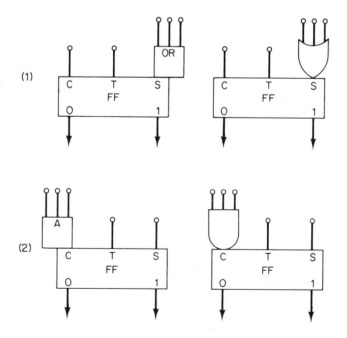

Fig. 6-23. Combined Functions.

example indicates that three *separate* signals are required to *clear* the flip flop. The AND and OR functions are part of the flip flop hardware.

6-4 BASIC LOGIC DIAGRAMS

We have seen most of the standard symbols. Now let's combine some symbols to determine how they are used to construct *logic diagrams.*

Double Gates

The double gate function has many variations and many uses. Figure 6-24 illustrates two AND gates *feeding* an OR gate.

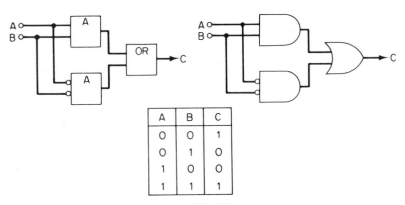

A	B	C
0	0	1
0	1	0
1	0	0
1	1	1

Fig. 6-24. Double Gate OR.

This OR function provides a high (1) output when *either* of the input AND gates is satisfied. In this case, both AND gates have inputs from the same source. When both A and B are high (1), the top AND is satisfied and it produces a high (1) for an input to the OR gate. When both inputs are low (0), the bottom AND is satisfied and it produces a high (1) for an input to the OR gate. So, C = AB, but also, C = $\bar{A}\cdot\bar{B}$. However, when we have one high (1) input and the other input is low (0), neither AND gate is satisfied, and both inputs to the OR gate are low (0). With two lows (0s) in, the OR gate is not satisfied, and the output at C is low (0).

By switching the state indicators (bubbles) among the legs of the AND and OR gates, and using more or less state indicators, we can create several variations of the double AND-OR gate.

Figure 6-25 shows a double gate with two OR gates feeding a NAND gate.

With this gating arrangement, the *inactive* output at C is a high (1). In order to obtain a low (0) output, we must satisfy *both* OR gates. When both A and B are low (0), OR gate 2 is satisfied, but OR 1 is not. This provides one

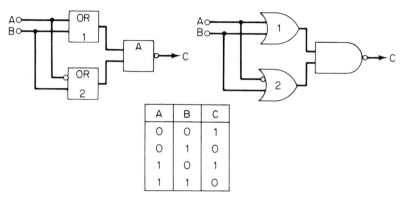

A	B	C
0	0	1
0	1	0
1	0	1
1	1	0

Fig. 6-25. Double Gate NAND.

high (1) and one low (0) input to the NAND gate, and the output remains high (1).

When we apply a high (1) at A and a low (0) at B, OR 1 is satisfied, but OR 2 is not. This also *fails* to activate the NAND gate, and the output remains high (1).

The remaining two input combinations (A = 0 and B = 1, or A = 1 and B = 1), will satisfy both OR gates. This places a high (1) on *both* legs of the NAND gate. The NAND gate is *activated*, and the resulting output becomes a low (0). Therefore, $\bar{C} = \bar{A}B + AB$. We read this as not C = not A and B OR A and B. Check this formula against the truth table.

Triple Gating

This time, we'll mix the gates and use four inputs. Consider the logic gates in Figure 6-26, and determine *all* the possible combinations that will provide an *activated* output at point E.

The final gate is a NAND, which gives this bit of logic an overall NAND function. The *inactive* output is a high (1), and in order to obtain a low (0) output, we must have a high (1) for input D and a low (0) from OR 2. So, D = 1 is one of our necessary conditions.

To obtain a low from OR 2, it must be *inactive*. The only way to make OR 2 inactive is to *activate* OR 1 and *deactivate* both AND gates. What combinations of ABC will accomplish this?

Let's try this one: $E = \overline{AB}CD$. The high (1) at D conditions one leg of AND 2. A low (0) at A conditions another leg of AND 2, but the low (0) at B deconditions one leg of both AND gates. So, both AND gates are off, with the resulting low (0) outputs. The low (0) at B deconditions one leg of OR 1, but the other leg is conditioned by the low at point C. OR 1 is *activated* and produces a high (1) output. The center leg of OR 2 *requires* a low (0) for activa-

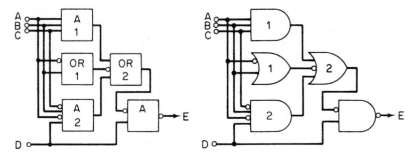

A	B	C	D	E
0	0	0	0	1
0	0	0	1	0
0	0	1	0	1
0	0	1	1	0
0	1	0	0	1
0	1	0	1	1
0	1	1	0	1
0	1	1	1	0
1	0	0	0	1
1	0	0	1	1
1	0	1	0	1
1	0	1	1	1
1	1	0	0	1
1	1	0	1	0
1	1	1	0	1
1	1	1	1	1

Fig. 6-26. Triple Gate NAND.

tion; so, the high (1) input deactivates this leg. The remaining two legs are deactivated by low (0) inputs from the deactivated AND gates. OR 2 remains deactivated with the resulting low (0) output.

The NAND gate has *one* leg conditioned by the high (1) at D, and the *other* conditioned by the low (0) from OR 2. The NAND gate is activated, and it produces a low (0) output at point E.

Let's try another combination: E = ĀBCD. We will take a few shortcuts on this one. The low (0) at A deconditions AND 1 and conditions OR 1. The high (1) at C deconditions AND 2. This provides the three *improper* inputs for OR 2 and leaves us with a low (0) output. The NAND gate is activated, and our output is low (0).

According to the truth table, there are *two other* combinations that will give us a zero output. They are: E = $\overline{AB}$CD and E = AB$\overline{C}$D. *Trace* these two conditions and *verify* the accuracy of the truth table. Trace a few *other* combinations to verify that *no* others will satisfy the conditions for a zero output.

6-5 EXTENDED USE OF LEVEL INDICATORS

By this time, you should be feeling fairly comfortable with *positive* logic. It is reasonably simple to interpret *either* positive or negative logic as long as you are sure which of the two you are trying to read. In many large systems, we find that the central computer is represented by positive logic, and one (or more) of the input/output devices is depicted with negative logic. Where the two types of logic join is an area of *hybrid* logic. Unfortunately, this is an area of great confusion to many people, but there is no need for you to be among the confused. The key is a thorough understanding of the use of state indicators.

Logic Identity

Figure 6-27 shows examples of proper identity for positive logic without electric inversion.

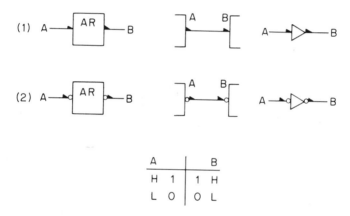

A		B	
H	1	1	H
L	0	0	L

Fig. 6-27. Identifying Positive Logic.

The table is both a table of combinations (low and high) and a truth table (0 & 1).

The *solid* flags tell us that the relative levels are: 1 = high and 0 = low; this is positive logic *identification*. Input and output level indicators are identical, which shows that we have positive logic on both sides of the logic symbol. On the connecting line between logic functions, we have solid flag indicators on both ends of the line. This shows that we have positive logic for the length of the line.

In example 2, we have bubble indicators mixed with our flag indicators. These tell us the relative level of the *activated* inputs and outputs. When the the input to the amplifier is high (1) at flag A, it provides a low (0) to activate the amplifier. The activated output of the amplifier is a low (0) which becomes a high (1) at point B.

On the connecting line between functions in example 2, a low (0) output gives us a high (1) at A and B which becomes a low (0) input to the next logic function.

Figure 6-28 shows identification of *negative* logic without electric inversion.

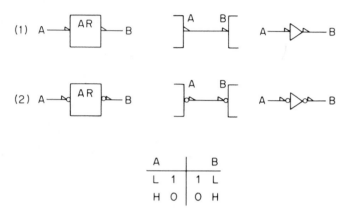

<table>
<tr><td>A</td><td>B</td></tr>
<tr><td>L 1</td><td>1 L</td></tr>
<tr><td>H 0</td><td>0 H</td></tr>
</table>

Fig. 6-28. Identifying Negative Logic.

The *open* flag identifies negative logic where the relative high is a 0 and the relative low is a 1. The amplifiers have identical flags on input and output, which shows negative logic on both sides of these symbols. The connecting lines have open flags on both ends to indicate negative logic for the length of the line. The bubble indicators show *activated* levels and have no bearing on the logic identity.

In example 2, the amplifier is activated by a low (1) and provides a low (1) activated output. In order to activate the amplifier, we need a high (0) at A, and the activated output again becomes a high (0) at flag B.

Logic Negation

Now we are in the *fringe area* between positive and negative logic. There is no electric inversion, but we have an abrupt *switch* from one type of logic to another. This is our area of hybrid logic. The change from positive to negative logic is illustrated in Figure 6-29.

In all cases in these two examples, the relative levels are reversed between A and B. The *solid* flag at A indicates that 1 is a relative high. The *open* flag at B indicates that 1 is a relative low.

In example 1, a high (1) into the amplifier gives us a low (1) at point B. The logic negation takes place on the *output* of the amplifier. The connecting

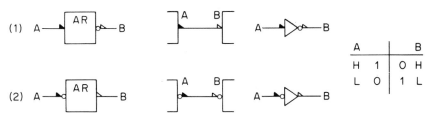

Fig. 6-29. Positive to Negative Negation.

lines between logic symbols have positive indicators on one end *and* negative on the other. This means that the logic negation takes place somewhere *along* the line.

In example 2, a high (1) into the amplifier provides a low (0) *activating* signal. This results in a low (1) output at B. The logic negation occurs on the *input* of this amplifier.

Figure 6-30 illustrates the transition from negative to positive logic. Again, there is *no* electric inversion, just a change in relative levels of one and zero.

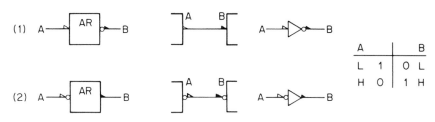

Fig. 6-30. Negative to Positive Negation.

The *open* flags at A indicate negative logic, where 1 is a relative low and 0 is a relative high. At point A, we have negative logic in each case. Point B, in each case, is shown as *solid* flags, for positive logic.

In example 1, the logic negation takes place on the *output* of the amplifier. In the connecting line, we have negative logic on one end and positive on the other. The negation is accomplished somewhere *along* the line. In example 2, the amplifier has negation on the input.

Logic Identity and Electric Inversion

Figure 6-31 illustrates several ways to provide positive logic identity and electric inversion at the *same* time.

The *solid* flags indicate positive logic at *all* points despite the fact that we have electric inversion in each logic function. A high (1) at A gives us a low (0) at B in every case, or a low (0) at A gives a high (1) at B.

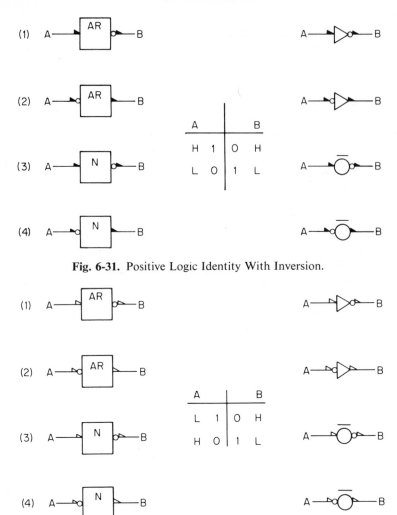

Fig. 6-31. Positive Logic Identity With Inversion.

Fig. 6-32. Negative Logic Identity With Inversion.

Figure 6-32 illustrates negative logic identity *combined* with electric inversion.

The *open* flags indicate that we have negative logic at *all* points, but at the same time, we have electric inversion in each function. A low (1) at any A gives us a high (0) at the corresponding B, or a high (0) at A gives us a low (1) at B.

Logic Negation and Electric Inversion

Figure 6-33 shows four examples of *hybrid* logic. Each example shows the *transition* from positive to negative logic *combined* with electric inversion.

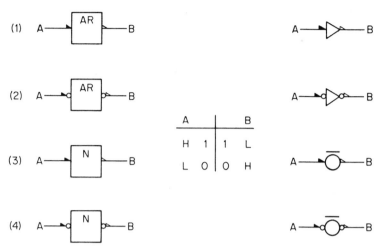

Fig. 6-33. Positive to Negative Negation With Inversion.

In each example, we have positive logic at point A identified by solid flags. Each function provides electric inversion, and each point B is negative logic. When we have a high (1) at any point A, it is inverted and becomes a low (1) at point B. Also, a low (0) at any point A is inverted and becomes a high (0) at point B.

Figure 6-34 shows another four examples of hybrid logic. This time, each example illustrates a negative to positive logic *transition* combined with electric inversion.

The open flags tell us that we have negative logic at all points A. The solid flags show that the logic is positive again at points B. Between these two points, we have both logic negation and electric inversion. A low (1) at any point A becomes a high (1) at the corresponding point B. Also a high (0) at any point A becomes a low (0) at the corresponding point B.

Logic Identity and Negation Between Functions

Figure 6-35 shows four examples of logic *identity* between level indicators. Each example also includes logic *negation* between the logic symbols. Please note that this is logic negation and *not* electric inversion. We have a table in each example which is both a table of combinations (H & L) and a truth table (1 & 0).

Examples 1 and 2 show positive logic *between* flag indicators. We know

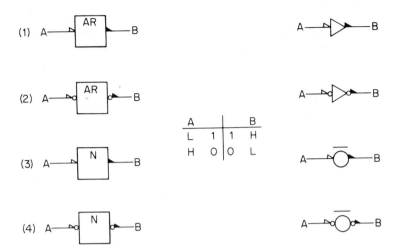

Fig. 6-34. Negative to Positive Negation With Inversion.

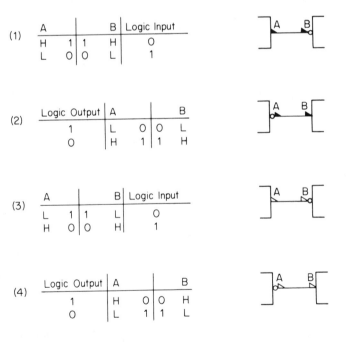

Fig. 6-35. Identity Between Indicators and Negation Between Functions.

this because both flags (A & B) are solid which not only shows that A is always identical to B, but also that a one is a relative high and a zero is a relative low. This fact is also specified in the table for *double* confirmation.

In example 1, when A is high (1), B is high (1); when A is low (0), B is also low (0). On the input to the next logic symbol, we have a bubble level indicator which signifies logic inversion. Thus, a high (1) at B will provide a 0 input to the next logic symbol. Notice that the table does not specify whether this input is relatively high or low. When going to *positive* logic, it is *low*; when going to *negative* logic, it is *high*.

In example 2, we are still in positive logic from A to B, and A and B are always identical. This time, our bubble is on the output of the last logic symbol. When our output is 1, A is low (0) and B is low (0). When we have a 0 output, A and B are high (1). The relative levels of the 1 and 0 output are governed by the *type* of logic in the *previous* stage.

Examples 3 and 4 show negative logic between flag indicators, and again we have logic negation between logic symbols. The open flags show negative logic by telling us that the 0 is relatively high. Again A and B are always identical. Check these facts in the tables.

In example 3, when A is high (0), B is high (0) and vice versa. When we have a high (0) at B, we have a 0 input to the next logic symbol. This is indicated by the bubble between flag B and the logic symbol. No effort has been made to show the relative levels of the inputs because they vary according to the type of logic in the next symbol. When we have a low (1) at A and B, the input to the next stage is a 1.

In example 4, we have a logic negation on the output of the previous logic symbol. When the output is 1, we have a high (0) at both A and B. When the output is zero, we have a low (1) at both A and B. The relative levels of the outputs are *dictated* by the type of logic in the previous stage.

Figure 6-36 shows four examples of logic *identity* between logic functions and logic *negation* between level indicators.

In examples 1 and 2 we have positive logic at point A which has become negative logic at point B. When A is a high (1), B is a high (0).

In examples 3 and 4, we have the opposite situation: negative logic at point A and positive logic at point B. When we have a high (0) at A, we have a high (1) at B.

Mixed Examples

Figure 6-37 consists of numerous uses of the flag indicator to depict *activating* inputs and *activated* outputs. No effort is made to separate positive and negative logic. The table of combinations is based on the assumption that the relative high is a one.

On the top row, we have an AND function that is activated by two *high*

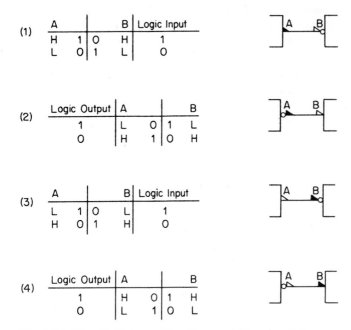

Fig. 6-36. Identity Between Functions and Negation Between Indicators.

inputs, and it produces a high output when activated. The OR function in this row is activated by any combination of *low* inputs, and its activated output is a low.

In the *fifth row* from the top, we have the reverse situation. Here the AND is activated by two *low* inputs, and the activated output is a low. The OR is now activated by any combination of *high* inputs, and its activated output is a high.

Study the other examples and verify the accuracy of the table of combinations.

Simplifying Level Indications

When both flag and bubble indicators appear on the same line, the bubble can be *eliminated* by replacing the flag with its *opposite* kind. This is illustrated in Figure 6-38.

In both examples, we show four ways to represent exactly the same thing. A, B, C, and D of example 1 show a high output from the previous function and a low input to the next function. In example 2, A, B, C, and D each shows a low output from the previous function and a low input to the next function.

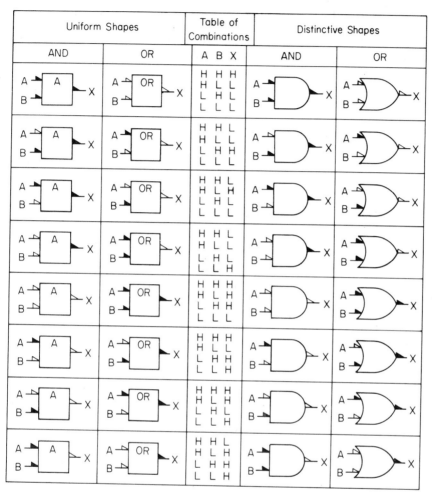

Fig. 6-37. Mixed Positive and Negative Logic.

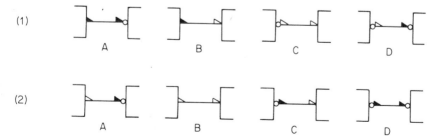

Fig. 6-38. Interpreting Level Indicators.

Magnetic Heads

The symbology for magnetic heads is frequently encountered. The symbols for these functions are illustrated in Figure 6-39.

WRITE   READ   ERASE   READ
                                      &
                                      WRITE

Fig. 6-39. Magnetic Heads.

Logic representation of a magnetic drum with magnetic heads may appear similar to the drawing in Figure 6-40.

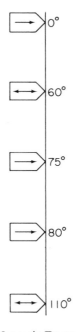

Fig. 6-40. Magnetic Drum With Heads.

6-6 DETAILED LOGIC DIAGRAMS

No matter how complex a sheet of logic may appear at first glance, you are *not likely* to encounter many symbols other than those we have *already* discussed. Figure 6-41 is a sample sheet of detailed positive logic using uniform shapes. Figure 6-42 is a sample sheet of the same logic using distinctive

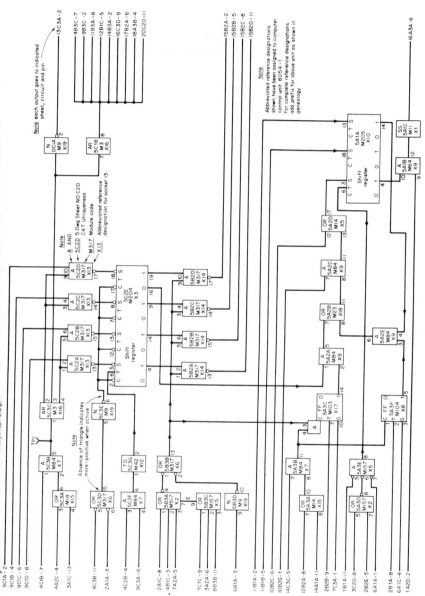

Fig. 6-41. Sample Standard Shape Logic.

151

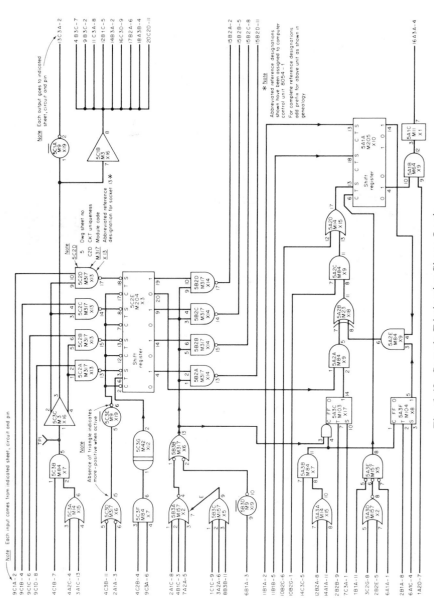

Fig. 6-42. Sample Distinctive Shape Logic.

152

shapes. We are *not* expected to interpret all the logic action because *insufficient* information is given. But you should be able to *identify* each function and specify the high or low activating and activated levels.

Review Exercises

1. Why is it necessary to be familiar with two systems of logic symbols?

2. Is it necessary to have a one to one ratio between logic symbols? Explain.

3. What is a logic function? Give three examples.

4. Distinguish between positive logic and negative logic.

5. In a system which has logic levels of +4 V and −2 V, what type of logic is used when the:
(a) +4 V level is a one?
(b) −2 V level is a one?

6. Using the levels in item 5, construct a table of combinations for a two legged device that satisfies the function of either a positive AND or a negative OR.

7. (a) What function is indicated by the truth table in Figure 6-43?
(b) Draw the symbol for this function in both popular systems of symbols.

A	B	C
0	0	0
0	1	0
1	0	0
1	1	1

Fig. 6-43. Truth Table for What Function?

8. (a) What function is represented by the symbol in Figure 6-44?
(b) Assume positive logic and construct a truth table for this function.

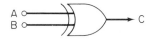

Fig. 6-44. Name the Function.

9. What is the formula for D in Figure 6-45?

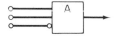

Fig. 6-45. State the Formula for D.

10. Draw the symbol for each of these functions using distinctive shapes:
    (a) $C = A + B$
    (b) $D = \overline{ABC}$
    (c) $C = \bar{A}B + A\bar{B}$
    (d) $\bar{D} = ABC$
    (e) $\bar{E} = A + B + C + \bar{D}$

11. Write the activated output formula for each of the functions in Figure 6-46.

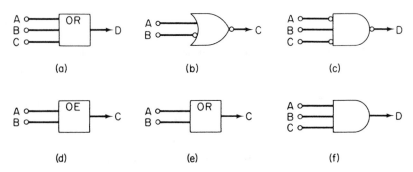

(a)   (b)   (c)

(d)   (e)   (f)

Fig. 6-46. Write the Output Formulas.

12. (a) What is the name of the logic function which provides a high out for a low in and vice versa?
    (b) Draw the logic symbol for this function in both systems of symbols.

13. Explain the meaning of the symbols in Figure 6-47.

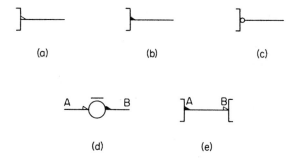

(a)   (b)   (c)

(d)   (e)

Fig. 6-47. Interpret the Symbols.

14. Construct a combined table of combinations and truth table for item 13d.

15. Draw the symbol for a two legged NAND gate in both systems of symbols, and construct the truth table for this function.

16. (a) Draw the symbol for a flip flop.

(b) Write the name for each of the three inputs and both outputs.

17. Explain what must be done to a flip flop (assume positive logic) to:

(a) store a one.

(b) store a zero.

(c) complement the content.

18. Assuming positive logic with levels of 0 V and $+6$ V, what are the outputs from the zero side and one side of a flip flop when it is:

(a) set?

(b) clear?

19. What is the primary function of the flip flop?

20. State the two principal ways that a flip flop latch differs from a flip flop.

21. What happens to a flip flop latch when a high is applied to both inputs at the same time?

22. Draw the logic symbol for a single shot and indicate a 1 μs delay.

23. What is the maximum period of time that a one bit can be stored in the single shot described in item 22? Why?

24. Describe the function of a Schmitt trigger.

25. Draw a distinctive shape symbol for a time delay with outputs of 1, 2, and 3 μs.

26. Interpret the symbol in Figure 6-48.

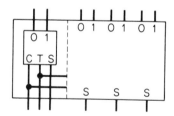

Fig. 6-48. What Does this Symbol Signify?

27. Draw a distinctive shape symbol showing two AND gates with a connector OR function on the output.

28. Draw a distinctive shape symbol to show that a flip flop is set by the output of a three legged AND gate which is part of the flip hardware.

29. Using standard shapes, draw a double gate that will perform the function of $C = AB + AB$.

30. Write the formula for the activated output of the circuit in Figure 6-49.

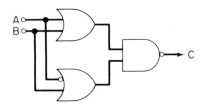

Fig. 6-49. Write Formula for Activated Output.

31. What type of logic is indicated with each symbol in Figure 6-50?

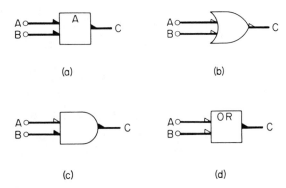

Fig. 6-50. Identify Types of Logic.

32. Write the formula for the activated output for each symbol in Figure 6-50.

33. Draw the logic symbol for each of the four types of magnetic heads, and indicate the function of each symbol.

logic
design

In 1854 an English mathematician, George Boole, published a book entitled *An Investigation of the Laws of Thought, on Which are Founded the Mathematical Theories of Logic and Probabilities*. It was his intention to perform a mathematical analysis of the *reasoning process*. Working with the laws of thought, he constructed a *logical* algebra which we call *Boolean* algebra.

Boolean algebra was first applied to electric circuits in 1938 as a *symbolic* analysis of relays and switching circuits. Since there is a close *relation* between the actions of a relay and the functions of digital circuits, we may apply the same basic techniques of design and maintenance to both.

The *variables* in computer circuits are always *one* and *zero*. These may also be represented by any two *opposite* conditions such as true or false, on or off, or open or closed. The variables used in Boolean equations have the same unique characteristics. They must always assume *one of two* possible values which we represent by a zero and one.

157

7-1 BASIC GATES AND CONVENTIONS

We have been *using* Boolean expressions since we first encountered the AND gate. Any equation which expresses the *relation* between input and output signals is a form of Boolean expression.

Connectives and Variables

We have already learned the use of the addition (+) sign and the multiplication (×) sign. The + means OR and the × means AND. Other signs or connections which in *conventional* mathematics indicate multiplication are also interpreted as AND functions. For example: AB, A·B, (A)(B), and A × B are all interpreted as A and B.

The *vinculum* (_____) serves a dual purpose. It is a symbol of *grouping* and a sign of *operation*. As a sign of operation, it indicates that the overlined term or terms are to be *complemented*. We refer to this as either complementation or *negation*. As a symbol of grouping, it *collects* the terms to be complemented. $\bar{A}$ + B shows that A is complemented. $\overline{A + B \cdot C}$ first collects the terms then complements.

Other signs of grouping are the parentheses, ( ), brackets, [ ], and braces, { }. These are used in the customary fashion to indicate that the *enclosed terms* are to be treated as a unit.

The equality sign (=) has the same meaning as in conventional mathematics.

Various letters of the alphabet are used to represent the *variables*. Since all variables are capable of assuming only one of two states, the numerals 0 and 1 are the only numbers needed in Boolean expressions.

Functions

Boolean algebra is primarily concerned with *three* functions: AND, OR, and NEGATION. These can be expanded to include our other two gates: NAND and NOR. This information should neither narrow your scope of thinking nor dull your enthusiasm for learning Boolean algebra. Even in early computers, these five functions represented from 50 to 75 percent of the circuits. In the new integrated machines, the percentage is even higher. With great redundancy of gates, it becomes economically important to keep the number to an absolute minimum. At first glance, we may assume that we need a combination of six gates to perform a particular task. After reducing the function to a Boolean expression, it can be analyzed and simplified. Instead of six gates, we may find that only one is needed. From the standpoint of economy, this is equivalent to the difference between a six transistor radio and a one transistor radio. When we multiply this by the several thousand gate combinations incorporated in a computer, the possible cost of nonessential gates borders on the fantastic.

OR Gate

The simple OR function is an important Boolean operation. The symbol and truth table are shown in Figure 7-1.

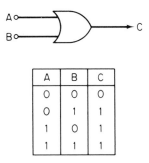

A	B	C
0	0	0
0	1	1
1	0	1
1	1	1

Fig. 7-1. OR Function.

The truth table tells us that the output(C) is a one when any or all inputs are ones. Thus,

$$0 + 0 = 0$$
$$0 + 1 = 1$$
$$1 + 0 = 1$$
$$1 + 1 = 1$$

Some texts refer to this as logical addition, but any effort to compare logical operations to conventional mathematics is an invitation to confusion. Let's simply read the plus sign as OR.

Our *variables* are A, B, and C, and any of these may be either a zero or a one. When C is the output and A and B are inputs, our Boolean formulas then become:

$$C = A + B$$
$$C = A + \bar{B}$$
$$C = \bar{A} + B$$

Notice that each formula is written so that the output is a one.

We may expand the OR gate to include any reasonable number of inputs. If we use six inputs, the output will be a one as long as we have a one on any input leg.

The OR gate is actually a *parallel* switching circuit as illustrated in Figure 7-2.

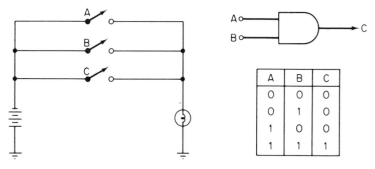

A	B	C
0	0	0
0	1	0
1	0	0
1	1	1

Fig. 7-2. Parallel Switching (OR)   **Fig. 7-3.** AND Function.
Circuit.

AND Gate

The AND gate in any specified type of logic is exactly *opposite* to the OR gate. We previously stated that one circuit could perform as either a negative AND or a positive OR. While the OR gives a one out for any one in, the AND gives a one out only when *all inputs* are ones. The AND symbol and truth table are shown in Figure 7-3.
We could rewrite the truth table like this:

$$0 \times 0 = 0$$
$$0 \times 1 = 0$$
$$1 \times 0 = 0$$
$$1 \times 1 = 1$$

This operation is frequently called logical multiplication, but this is another confusion factor. It is best that we consider all multiplication signs as AND functions.

Again the variables are A, B, and C but now *only one* combination will give us a one output. This is:

$$C = A \times B$$

This shows that the AND gate is a *series* switching circuit as illustrated in Figure 7-4.

Inverter Function

This is a function with a *single* variable. The inverter is a circuit which *complements* its input. The symbol and truth are shown in Figure 7-5.

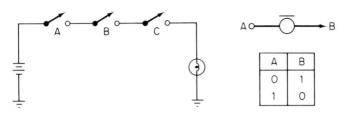

Fig. 7-4. Series Switching (AND)
Circuit.

Fig. 7-5. Inverter and
Truth Table.

When A is a one, B is NOT a one. This can be expressed as $\bar{B} = A$ or
$B = \bar{A}$.

In many cases, our inverter is combined with other logic symbols, and its
presence is indicated by our bubble state indicator. Some examples are shown
in Figure 7-6.

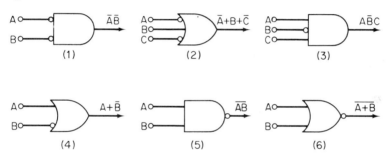

Fig. 7-6. State Indicators Showing Inversion.

In Figure 7-6, we have started indicating the output *in terms* of the inputs.
In examples 1, 2, 3, and 4, the inversion occurs on the *inputs*. Notice that in
these cases the individual outputs are *overlined* (NOTted). In examples 5 and
6, the inversion takes place in the *output*. Here the *entire* output expression,
including the signs of operation, are *overlined*. This distinction is very important because the complement (NOT) of $+$ becomes $\times$ and the complement of
$\times$ is $+$. Thus, $\overline{AB} = \bar{A} + \bar{B}$, and $\bar{A} + \bar{B} = \overline{AB}$. This is emphasized in
Figure 7-7.

In example 1, the inversion occurs *before* the variables are ANDed together.
This gives *each* variable in the output a *separate* NOT sigh. In examples 2 and
3, the inversion occurs *after* the inputs are combined (ANDed in 2 and ORed in
3). This makes it compulsory to NOT the *entire* output expression.

NAND Gate

The NAND gate is an AND function with an inverter on its output. The
symbol and truth table are shown in Figure 7-8.

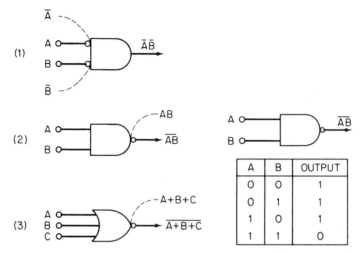

Fig. 7-7. NOTting Expressions.   **Fig. 7-8.** NAND Gate and
                                      Truth Table.

The *activated* output of the NAND gate is a zero, but our Boolean expressions are in terms of a *one* output. We have three combinations that give us
a one out; $\bar{A} \times \bar{B}$, $\bar{A} \times B$, and $A \times \bar{B}$. These are all summed up in the expression $\overline{AB}$ which means "the output is a one when either A or B is a zero."
We may rewrite the expression as A + B, but it does not adequately identify
the NAND gate.

NOR Gate

The NOR gate is an OR function with an inverter on the output. The symbol
and truth table are shown in Figure 7-9.

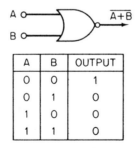

A	B	OUTPUT
0	0	1
0	1	0
1	0	0
1	1	0

Fig. 7-9. NOR Gate and Truth Table.

The *activated* output of the NOR gate is a *zero*, and *any* one input will
activate the gate. Our output expression is formed from the single *deactivated*
condition because this condition gives us a *one* output. In the truth table, we

see that this condition is both inputs zero ($\bar{A}$ and $\bar{B}$). The output expression $\overline{A + B}$ says that A and B were ORed together and then inverted. The expression could be rewritten as $\bar{A}\bar{B}$, but it would not adequately identify the NOR gate.

7-2 BOOLEAN EXPRESSIONS FROM LOGIC DIAGRAMS

We have carefully examined the *five* functions which are frequently described by Boolean expressions. When we carefully compose our output expression in terms of the inputs, the expression is a precise *description* of the logic symbol as well as its inputs.

In many cases, we find a particular signal processed through many orders of logic gates. By order, we mean that the output of one gate becomes the input to the next. The final output Boolean expression is a precise description of the *entire combination* of gates and the inputs to each gate.

First Order Logic

We have already covered *first order* logic; it is a single gate with simple inputs. The output expression tells us the *type* of gate, the *number* of legs, and the *inputs* that will give us a one output. For example: the expression A + B + C is the output of a three legged OR gate with inputs A, B, and C. When *either* A or B or C is a one, the output is a one. Examine the output expressions in Figure 7-10, and notice that each one is a precise description of the logic gate and its inputs.

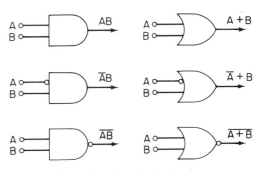

Fig. 7-10. First Order Logic.

Second Order Logic

When the output of one gate becomes the input to another, we are dealing with *second order* logic. This means that at least one input to second order logic is a combined Boolean expression describing the previous gate. Consider the OR gates in Figure 7-11.

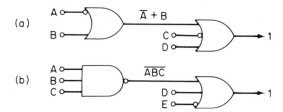

Fig. 7-11. Second Order Logic OR Gates.

In example a, our second order OR gate has two simple inputs (C and D) and one combined input from a first order OR gate ($\bar{A} + B$). Our problem is to compose an expression for the output at point 1. We must enclose the combined input in parentheses in the output expression. The output is ($\bar{A} + B$) + $\bar{C}$ + D. This says that our second order gate is an OR circuit with three legs. One leg has a simple input (D). One leg has a simple input, but it is inverted at the input ($\bar{C}$). The third leg has an input from a previous OR gate ($\bar{A} + B$).

In example b, we have a second order OR gate with two simple inputs (D and E) and one combined input from a first order NAND gate. At point 1, our expression is $\overline{ABC} + D + \bar{E}$. In this case, we *do not* enclose the combined input in parentheses. The variables in the combined input are grouped naturally by both ANDing and inversion; they cannot be confused.

Notice in each case that the final output expression is a precise description of both orders of logic.

Third Order Logic

When a second order logic output is used as the input to another gate, we are using *third order* logic. This is illustrated in Figure 7-12.

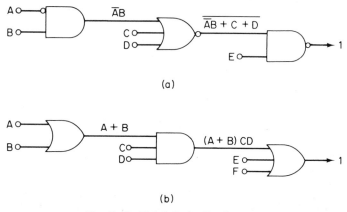

Fig. 7-12. Third Order Logic.

In example a, our third order logic is a NAND gate. The combined input accurately describes the two previous gates. We need to combine this with E and form an expression to describe all three gates. The combined expression must be enclosed because of the OR functions it contains. The expression at point one becomes $(\overline{\overline{\overline{A}B} + C + D})\overline{E}$. The double NOT sign $(=)$ shows that signals B, C, and D have been inverted twice, first by a NOR gate and then by a NAND gate. The triple NOT $(\equiv)$ shows that A has been inverted three times. The E has been inverted only once.

In example b, our third order logic is an OR gate. The combined input $(A + B)CD$ accurately describes the two previous gates. We need to combine it with E and F in a manner that describes all three gates. Since the combined input is ANDed, no further enclosures are needed. The output at 1 is $(A + B)CD + E + F$.

Fourth Order Logic

When the third order output is used as the input to another gate, we are dealing with *fourth order* logic. This is illustrated in Figure 7-13.

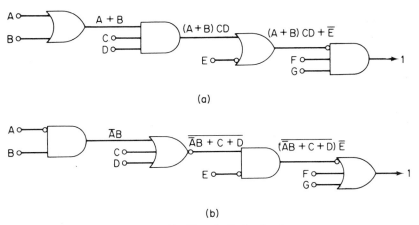

(a)

(b)

Fig. 7-13. Fourth Order Logic.

In both cases, our final output at point 1 must precisely describe all four gates and their inputs. In example a, the output is $[(A + B)CD + \overline{E}]FG$. In example b, the output is $(\overline{\overline{\overline{A}B} + C + D})\overline{E} + F + G$.

Subsequent Orders of Logic

There is no rule as to how many gates a signal may pass through. You may encounter *more than* four orders of logic. We have established the procedure for composing output expressions and this procedure will continue for any number of levels.

Before proceeding to new material, let's solve a few problems by way of review.

Problems

1. What is the proper Boolean expression at points 1, 2, 3, and 4 of Figure 7-14?

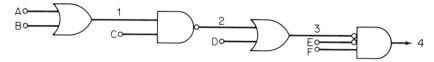

Fig. 7-14. Compose Boolean Expressions.

Point 1 is A + B
Point 2 is $\overline{(A + B)C}$
Point 3 is $\overline{\overline{(A + B)C} + D}$
Point 4 is $\overline{(A + B)(C + D)}\bar{E}F$

2. What is the proper Boolean expression at point 1 of Figure 7-15?

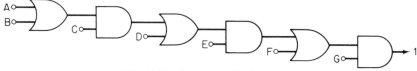

Fig. 7-15. Compose Final Output.

The proper expression at point 1 is {[(A + B)C + D]E + F}G.

If you made any errors in the solution of these two problems, you should review until all procedures are clear in your mind.

7-3 LOGIC DIAGRAMS FROM BOOLEAN EXPRESSIONS

Since a proper Boolean expression is an accurate description of the logic and the inputs, we should be able to translate final output Boolean expressions into logic diagrams.

First and Second Order Logic Expressions

Expressions representing first and second order logic offer little challenge because the complete diagram flashes to mind the instant we see the Boolean expression. For instant, the term ABC can be only the output from a first order, three legged AND gate with A and B and C as inputs. The term A + B can be produced only by a two legged, first order OR gate with inputs of A and B.

The term (A + B)(C + D) indicates a *second order* AND gate preceded by *two* first order OR gates. One of these OR gates has inputs A and B, and the inputs to the other are C and D.

Boolean to Logic Conversion

Drawing a logic diagram from a Boolean expression is a *skill*. We can learn the procedure from a book, but proficiency comes only with *practice*. Consider this Boolean expression: [(A + B + C)(D + E) + F + G(H + I)]J. It looks complicated, but it really isn't. Notice that *all* variables and *function* signs, except one, are *enclosed* in brackets. This should tell us the function of our final gate.

What did you conclude? Well, the J is ANDed with the enclosed expression. The *final* gate must be a *two* legged AND gate with J on one leg and everything else on the other. Draw the final gate showing inputs and output. Your drawing should *match* that in Figure 7-16.

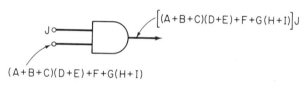

$$\left[(A+B+C)(D+E)+F+G(H+I)\right]J$$

(A+B+C)(D+E)+F+G(H+I)

Fig. 7-16. Final Gate with Inputs and Output.

On careful examination of the combined input, (A + B + C)(D + E) + F + G(H + I), we find *two* OR signs that *are not* enclosed. From this we know that the previous gate is a *three* legged OR. It has F on one leg, G(H + I) on another leg, and (A + B + C)(D + E) on the third leg. Draw this OR gate; combine it with the AND gate just completed; and show the inputs. The result should be logically the *same* as Figure 7-17.

We have *two* combined inputs to the OR gate. Each of these is supplied from a two legged AND circuit. Let's construct the AND gate which supplies

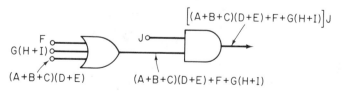

$$\left[(A+B+C)(D+E)+F+G(H+I)\right]J$$

F
G(H+I)
(A+B+C)(D+E)   (A+B+C)(D+E)+F+G(H+I)

Fig. 7-17. OR Gate Supplying the Combined Input.

the input G(H + I). It must be a simple AND gate with G on one leg and H + I on the other. Connect it into your diagram, and show the inputs. The result should *match* Figure 7-18.

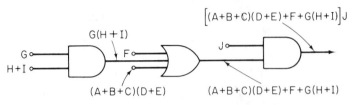

Fig. 7-18. AND Gate Supplying G(H + I).

We still have a combined input (H + I) indicating *another* gate to the left of this AND gate. It is an OR gate with H on one leg and I on the other. Draw it into your diagram and *compare* with Figure 7-19.

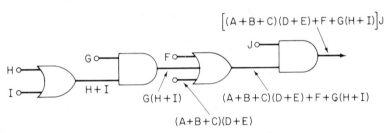

Fig. 7-19. Adding the First Order OR Gate.

Now we must *return* to our first OR circuit and consider the combined input on its third leg, (A + B + C)(D + E). We said that this was supplied by a *two* legged AND gate. This gate has A + B + C on one leg and D + E on the other. Draw it into your diagram and *compare* with Figure 7-20.

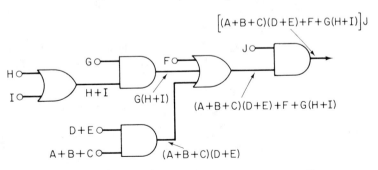

Fig. 7-20. AND Gate Supplying (A + B + C)(D + E).

Both inputs to this AND gate are *combined* inputs from OR circuits. One OR circuit has *two* legs with inputs D and E; the other has *three* legs with inputs A, B, and C. Draw in both of these OR circuits to complete your logic diagram. *Compare* the final results with Figure 7-21.

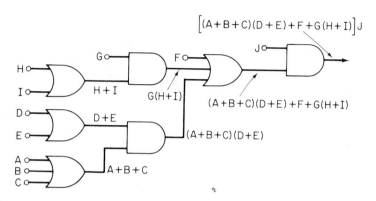

$$\left[(A+B+C)(D+E)+F+G(H+I)\right]J$$

Fig. 7-21. Final Logic Diagram.

A short review is in order. The output expression had *one* operation sign that was not enclosed. This was an AND sign which revealed the nature of our *output* gate. As we worked toward the lowest order of logic, we *removed* a sign of grouping as we passed from output to inputs. The *exposed* operation sign in each case revealed the nature of the *previous* gate. We drew in the gates one at a time until all signals were traced back to their *lowest* logic order.

Let's step our way through another conversion. Here is the Boolean expression: $(\overline{AB + C})D + [F(G + H) + I]$.

The expression contains *one* OR sign that *is not* enclosed. So, the final gate is an OR gate with two inputs. The variables to the *left* of the exposed OR sign form the input for one leg. The other leg has the combined variables to the *right* of the exposed OR sign. The OR gate with inputs and output is illustrated in Figure 7-22.

$$(\overline{AB+C})D + \left[F(G+H)+I\right]$$

Fig. 7-22. Final OR Gate with Inputs and Output.

Both inputs are fairly complex, so we will concentrate on one at a time. First the signal $(\overline{AB + C})D$. We have an *exposed* AND function which indicates a *two* legged AND gate with D on one leg and $\overline{AB + C}$ on the other. Placing this with our OR gate produces the diagram in Figure 7-23.

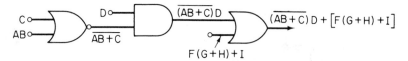

Fig. 7-23. Adding the AND Gate.

One leg of the AND gate shows a combination signal, AB + C. This started as two OR inputs to the *previous* stage. After they were ORed together, they were *inverted*. This identifies our *next* lower order logic symbol as a NOR gate. Adding it to our diagram produces Figure 7-24.

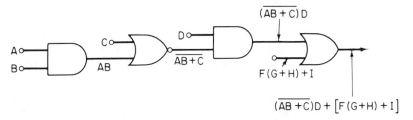

Fig. 7-24. NOR Gate which Produces $\overline{(AB + C)}$.

The AB input to the NOR gate indicates that this input comes from a two legged AND gate. It has A on one leg and B on the other. Let's draw it in and *compare* with Figure 7-25.

Fig. 7-25. The First Order Logic for Half of the Final Output.

We now have the logic which produces half of our final output. To this we must connect the logic which produces the signal F(G + H) + I. This signal has one exposed OR sign which indicates that it comes from a two legged OR gate. The OR gate has F(G + H) on one leg and I on the other. Adding this gate expands our logic diagram as shown in Figure 7-26.

Examining the combined input to our new OR gate, we see that it is an ANDed signal. It came from a two legged AND gate with F on one leg and G + H on the other. When this AND gate is added, our logic appears as in Figure 7-27.

This AND gate still has a combined input, G + H. This comes from a two legged OR circuit. Draw it in and compare with Figure 7-28.

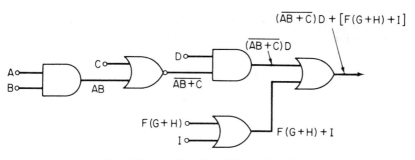

Fig. 7-26. OR Gate for F(G + H) + I.

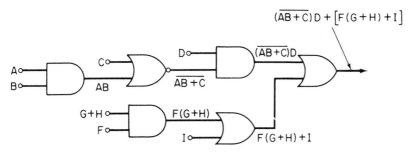

Fig. 7-27. AND Gate for F(G + H).

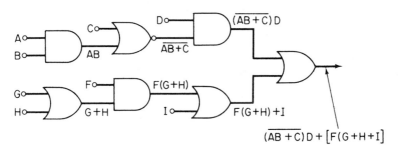

Fig. 7-28. Logic Diagram for (AB + C)D + [F(G + H) + I].

7-4 BOOLEAN LAWS

As we have already seen, any combination of *logic* gates can be *converted* to a *Boolean* expression, and any proper *Boolean* expression can be *converted* to a *logic* diagram. These two conversions are two of the most *useful tools* the computer technician has at his diaposal. But if we are concerned with logic design, a simple conversion is not enough. We must be able to *manipulate* the Boolean expression and obtain its *simplest* form. Reducing all Boolean expressions to their simplest form before converting them to logic will result in a *minimum* number of logic gates. This is an important *economic* consideration in computer design.

Results of Simplifying Expressions

To illustrate the importance of simplifying the Boolean expressions, we will show logic diagrams for expressions *before and after* simplification. Figure 7-29 is the logic diagram for the Boolean expression (ABC)(DEF + $\overline{DEF}$).

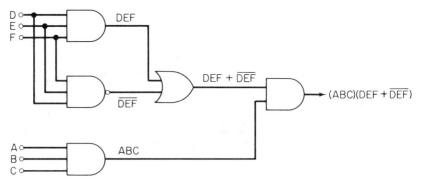

Fig. 7-29. Logic Diagram for (ABC)(DEF + $\overline{DEF}$).

The simplest form of the Boolean expression in Figure 7-29 is ABC. The logic diagram for this simplified expression is a single AND gate as shown in Figure 7-30.

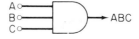

Fig. 7-30. Logic Equivalent of Figure 7-29.

Logically there is no difference between the expressions (ABC)(DEF + $\overline{DEF}$) and ABC. Consequently, the logic diagram constructed from either equation will do the same job. However, in this case, we can save one OR gate, one NAND gate, and two AND gates by simplifying the expression.

Here is another example. The logic diagram for the Boolean expression (A + B)(B + C) + ($\overline{A + B}$)($\overline{B + C}$) is shown in Figure 7-31.

The Boolean expression that we used to construct the logic diagram in Figure 7-31 can be simplified. In its simplest form it is AC + B + AC. When we translate this into a logic diagram it appears as in Figure 7-32.

The logic diagram in Figure 7-32 will do the *same* job as that in Figure 7-32, but it does so with a *saving* of two OR gates and two NOR gates.

7-5 RULES FOR SIMPLIFYING EXPRESSIONS

There are *ten* basic Boolean laws and *two* common identities which govern the simplification of Boolean expressions. Let's examine each law briefly then apply them to simplify some expressions.

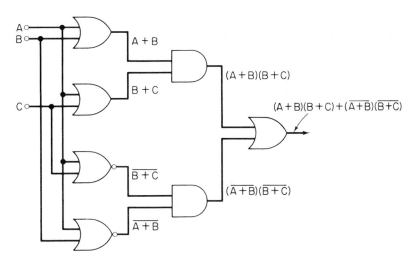

Fig. 7-31. Logic Diagram for $(A + B)(B + C) + \overline{(A + B)}$ $\overline{(B + C)}$

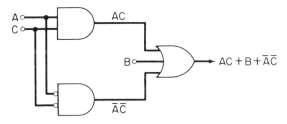

Fig. 7-32. Logic Equivalent of Figure 7-31.

Law of Identity

The law of identity states that any letter, number, or expression is equal to itself. Here are some examples:

$$A = A$$

$$\bar{A} = \bar{A}$$

$$1 = 1$$

$$AB = AB$$

Commutative Law

The commutative law specifies that the order of the AND and OR connectives does not affect the results. Some examples are:

$$AB = BA$$
$$A + B = B + A$$
$$ABC = BCA = CBA$$
$$R(S + T) = (T + S)R$$

Law of Association

The association law states that the variables of an expression may be grouped in any quantity as long as they are connected by the same sign.

$$A(BC) = ABC$$
$$A + (B + C) = A + B + C$$
$$(AB)(CD)(EF) = ABCDEF$$

Idempotent Law

The idempotent law states that when a term is ANDed or ORed with itself, the result is equal to the original term: $AA = A$ and $A + A = A$. This is illustrated in Figure 7-33.

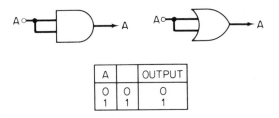

A		OUTPUT
0	0	0
1	1	1

Fig. 7-33. Idempotent Law.

Law of Involution

The law of *involution* is also called the *double negation* law and the law of *complementation*. $\bar{1}$ (NOT 1) is equal to 0. We know this is true because we have only two states, so if it is *not one* it must be zero. $\bar{0}$ (NOT 0) is equal to one, and when $A = 1$, $\bar{A} = 0$. These terms are complements of one another.

The logic connectives are also complemented. A *complemented* OR sign, $\mp$, is equivalent to an AND sign, x. A *complemented* AND sign, $\bar{x}$ is equivalent to an OR sign, $+$. Thus, $\overline{A + B} = \bar{A} \cdot \bar{B}$ and $\overline{AB} = \bar{A} + \bar{B}$.

We frequently find expressions with a double negation sign like this $\overline{\overline{A}}$. Any expression that is complemented *twice* is equal to the original expression: $\overline{\overline{A}} = A$, $\overline{\overline{1}} = 1$, and $\overline{\overline{0}} = 0$. It is not uncommon to find an expression with a *triple* negation sign, $\overline{\overline{\overline{A}}}$. This is equivalent to one negation. We may generalize by saying any *even* number of complements gives us the *original* expression, and any *odd* number of complements is the equivalent of complementing *one time*.

Law of Intersection

The law of intersection tells us that any expression ANDed with zero is equal to zero, and any expression ANDed with one is equal to the original expression.

$$A \times 0 = 0$$
$$A \times 1 = A$$

Law of Union

The law of union states that when we OR an expression with zero, we obtain the original expression, and when we OR an expression with one, we obtain a one.

$$A + 0 = A$$
$$A + 1 = 1$$

DeMorgan's Theorems

$$\text{First:} \quad \overline{A + B} = \overline{A} \times \overline{B}$$
$$\text{Second:} \quad \overline{AB} = \overline{A} + \overline{B}$$

These two theorems establish our procedures for *splitting* and *joining* vincula. When we *split* a vinculum such as $\overline{A + B}$, the function sign is changed. In the first theorem, $\overline{A + B}$ becomes $\overline{A} \times \overline{B}$. In the second theorem, $\overline{AB}$ becomes $\overline{A} + \overline{B}$.

We can *reverse* the theorems as follows without changing the meaning.

$$\text{First:} \quad \overline{A} \times \overline{B} = \overline{A + B}$$
$$\text{Second:} \quad \overline{A} + \overline{B} = \overline{AB}$$

This shows that the rule for joining is the *same* as the rule for splitting vincula.

Remember that the vinculum is a NOT sign as well as a sign of grouping. The rules previously discussed concerning *multiple* complements also apply to the function signs. That is, when a sign is complemented any even number of times, it is equivalent to the original sign. When a sign is complemented any odd number of times, the AND becomes an OR and the OR becomes an AND.

Distributive Law

The distributive law is in two parts:

$$\text{Part 1:} \quad A(B + C) = AB + AC$$
$$\text{Part 2:} \quad A + BC = (A + B)(A + C)$$

In some cases, the form of an expression must be altered before it can be simplified. Examine the expression on each side of the equality sign in both parts of this law. In some cases, the left hand expression is more convenient. At other times, we need the right hand expression. We can apply this law to convert in either direction.

When changing from *left to right*, we multiply as in algebraic *multiplication* of a polynomial by a monomial.

$$A(B + C) = AB + AC$$
$$A + BC = (A + B)(A + C)$$

When changing from *right to left*, the procedure is similar to *factoring*.

$$AB + AC = A(B + C)$$
$$(A + B)(A + C) = A + BC$$

We may apply the distributive law to only *part* of an expression when we find it convenient to do so. For instance, $(A + B)(C + B) + (\overline{C + B})$ $(\overline{A + B})$ can be changed to $B + AC + (\overline{C + B})(\overline{A + B})$. We must be *careful* when altering part of an expression. A *good test* is to place brackets around the part we plan to alter. If the brackets *do not* change the meaning, we may apply the law.

Law of Absorption

The law of absorption also deals with *factoring*. In this case, we are factoring a letter from a term which contains only that letter. When this is done, the factored letter is replaced by a one.

$$A + AB =$$
$$A(1 + B) =$$
$$(A \times 1) + (AB) =$$
$$A$$
$$D + CE = D(1 + E) = D \times 1 = D$$
$$\text{Also, } A(A \times B) = A$$

Common Identities

Two common identities have been *derived* from the basic Boolean laws, and they help us to perform *rapid* simplification of an expression.

$$\text{First:} \quad A(\bar{A} + B) = AB$$
$$\text{Second:} \quad A + \bar{A}B = A + B$$

Summary of Basic Laws

The ten basic laws are repeated here for quick reference.

Identity: $A = A; \bar{A} = \bar{A}$

Commutative: $AB = BA; A + B = B + A$

Associative: $A(BC) = ABC; A + (B + C) = A + B + C$

Idempotent: $AA = A; A + A = A$

Involution: $\bar{\bar{A}} = A; A\bar{A} = 0; A + \bar{A} = 1$

Intersection: $A \times 1 = A; A \times 0 = 0$

Union: $A + 1 = 1; A + 0 = A$

DeMorgan: $\overline{AB} = \bar{A} + \bar{B}; \overline{A + B} = \bar{A}\bar{B}$

Distributive: $A(B + C) = AB + AC; A + BC = (A + B)(A + C)$

Absorption: $A(A + B) = A; \quad A + AB = A$

Common Identities: $A(\bar{A} + B) = AB; A + \bar{A}B = A + B$

7-5 APPLICATION OF BOOLEAN LAWS

We will *simplify* one expression and identify the law used on each step. Then we will work out several problems and allow you to determine which of the laws are being applied.

Example

$(A + B)(C + B) + (\overline{A + B})(\overline{C + B})$	
$AC + AB + BC + BB + (\overline{A + B})(\overline{C + B})$	Distributive
$B + AB + BC + AC + (\overline{A + B})(\overline{C + B})$	Idempotent-Associative
$B + AC + (\overline{A + B})(\overline{C + B})$	Absorption
$B + AC + \bar{A}\bar{B}\bar{C}$	Distributive-DeMorgan's
$AC + B + \bar{A}\bar{B}\bar{C}$	Commutative
$AC + (B + \bar{A})(B + \bar{B})(B + \bar{C})$	Distributive
$AC + (B + \bar{A})(1)(B + \bar{C})$	Complementary
$AC + (B + \bar{A})(B + \bar{C})$	Intersection
$AC + BB + B\bar{C} + \bar{A}B + \bar{A}\bar{C}$	Distributive
$AC + B + B\bar{C} + \bar{A}B + \bar{A}\bar{C}$	Idempotent
$AC + B + \bar{A}\bar{C}$	Absorption

Problems

(1) $(RPQ + \overline{QRP})(ADE)$

Solutions

$RPQ + \overline{RPQ} = 1$

$1 \times ADE = ADE$

(2) $A + \bar{A}B$

$A + \bar{A}B = (A + \bar{A})(A + B)$
$A + \bar{A} = 1$
$1(A + B) = A + B$

(3) $A(\bar{A} + B)$

$A(\bar{A} + B) = A\bar{A} + AB$
$A\bar{A} = 0$
$0 + AB = AB$

(4) $(\overline{XY + Z}) + (XY + Z)WT$

$(\overline{XY + Z}) + (XY + Z)WT =$
$(\overline{XY + Z} + XY + Z)$
$(\overline{XY + Z} + WT)$
$(\overline{XY + Z} + XY + Z) = 1$
$(1)(\overline{XY + Z} + WT) =$
$(\overline{XY + Z} + WT) =$
$(\bar{X} + \bar{Y})\bar{Z} + WT =$
$\bar{Z}\bar{X} + \bar{Y}\bar{Z} + WT$

7-6 VEITCH DIAGRAMS

Diagrams may be used to *aid* in the simplification of Boolean expressions. There are several variations of these diagrams, but all perform basically the same function. As a group, we refer to them as *logic maps* or truth diagrams. Some of the names are Venn diagrams, Karnaugh maps, and *Veitch* diagrams. The latter were named for Edward Veitch of the Burroughs Corporation. Veitch presented chart methods to indicate logic statements concerning computer circuits in 1952. We will use a variation of the Veitch diagrams. Diagrams become essential when we encounter expressions that cannot be minimized by the basic Boolean laws. Here is an example that might cause problems: $AB + \bar{B}\bar{C} + A\bar{C}$. The simplest form for this expression is $AB + \bar{B}\bar{C}$, but it is difficult to obtain by the basic laws alone. It is easy when plotted on a Veitch diagram.

Minterm

A minterm is the *symbolic* product of a given number of variables. Suppose that we have the *three* variables; R, S, and T. All the following expressions are minterms of these three variables: $TR\bar{S}$, $RS\bar{T}$, AND $\bar{R}\bar{T}\bar{S}$. For the *four* variables, W, X, Y, and Z, we have these minterms: $W\bar{X}Y\bar{Z}$, $WXYZ$, and $\bar{Z}W\bar{Y}\bar{X}$. We must reduce our expression to *minterm* form before we can plot the variables on a Veitch diagram.

Minterm Type Term

A minterm *type* term is a minterm with one or more *variables missing*. The following expressions are minterm *type* terms for the variables, J, K, L, and M. JKL, K, LM, and MJK. When our expression is in minterm form, it is composed *entirely* of minterms, minterm type terms, or a mixture of the two. A minterm expression has *no part* enclosed by grouping signs and no vinculum that covers more than one letter.

Converting to Minterm Form

There are *three* steps to converting an expression to minterm form. They are:

1. Split or remove vincula.
2. Remove signs of grouping.
3. Simplify within the term.

Convert this term: $\overline{T + V}$ + RS.
    Solution: $\overline{T}\overline{V}$ + RS. Stop here; this is in minterm form.
        Try this one: $\overline{\overline{AB}}$ + CD + $\overline{\overline{E}B}$ + $\overline{\overline{BCD}}$ + $\overline{\overline{E}}$
    Solution: AB + CD + EB + BC + $\overline{D}$ + $\overline{E}$. We could simplify further, but this is our minterm form.

Determining the Number of Minterms

Before we can plot our variables, we must determine how many variables we have and the possible number of minterms. For *three* variables, we have *eight minterms;* for *four* variables, we have *16 minterms.* After we count our variables, we can use that number as a *power of two* to determine our possible number of minterms. For the expression, J + KL + MN, we have five variables and 2^5, or 32, minterms.

Constructing the Diagram

After we have determined the number of minterms in an expression, we draw a diagram with *one square* for each minterm. This is illustrated in Figure 7-34.

Part a of Figure 7-34 has a square for each of the *eight* minterms in a *three* variable expression. A *four* variable expression has *16 minterms* and requires a 16 *square* diagram as in part b. Part c illustrates a *32 square* diagram for a *five* variable expression.

Plotting Three Variables

Figure 7-35 illustrates the method of plotting a *three* variable expression.
Our three variables are A, B, and C. Notice that we use each of these and the complement of each. A close examination of the diagram will reveal that

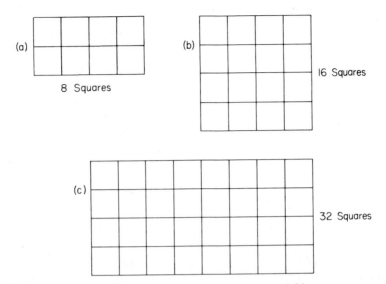

Fig. 7-34. Diagrams for 3, 4, and 5 Variables.

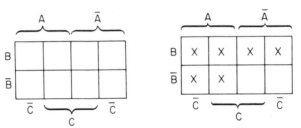

Fig. 7-35. Assigning Squares.

Fig. 7-36. Veitch Diagram for A + B.

half of the squares are assigned to *each* of the variables, and the other half are assigned to the *complement* of each variable. Each variable overlaps *every other* variable and all complements *except* its own.

Figure 7-36 is a plot of the expression A + B.

First, we place an x in each square that is assigned to A; then we place an x in every square that is assigned to B. That's simple enough, isn't it? Let's try a few more plots to make sure that we have the procedure. Then we will start extracting the simplest expression.

When our terms are other than A, B, and C it is recommended that the format (alphabetical order) of Figure 7-36 be followed for best comparison of your work with the examples and answers given. Plot B + C + Ā. Your plot should match Figure 7-37.

The past two examples illustrate that a *one* variable term, such as A, occupies *four* squares. As the number of variables in a term *increase*, the num

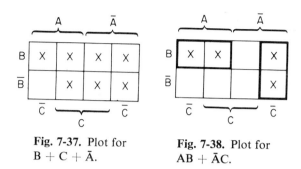

Fig. 7-37. Plot for
B + C + Ā.

Fig. 7-38. Plot for
AB + ĀC.

ber of squares *decrease*. A *two* variable term, such as AB, occupied *two* squares, and a *three* variable term, such as ABC, occupies *only one* square. Figure 7-38 is a plot of two terms with two variables in each term.

The procedure with a *two* variable term is to plot only the squares that are *common* to both variables. Plot this expression: EF + FĒ. Check your results against Figure 7-39.

Plot this expression: C̄B + AC̄ + B̄C + B̄Ā. Compare your results with Figure 7-40.

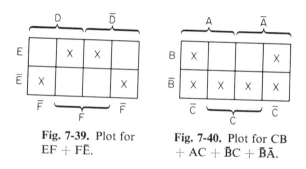

Fig. 7-39. Plot for
EF + FĒ.

Fig. 7-40. Plot for CB
+ AC + B̄C + B̄Ā.

When we go to the *three* variable term, we continue the practice of plotting *only* the squares that are *common* to all variables in the term. This is illustrated in Figure 7-41.

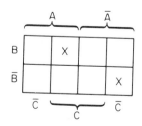

Fig. 7-41. Plot for ABC + ĀB̄C.

We have one square that is common to A, B, and C, and one square that is common to Ā, B̄, and C̄.

Plot the expression $\overline{W}\overline{X}Y + W\overline{X}Y + \overline{W}X\overline{Y}$. Check your results against Figure 7-42.

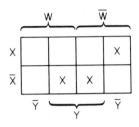

Fig. 7-42. Plot for $\overline{W}\overline{X}Y + W\overline{X}Y + \overline{W}X\overline{Y}$.

Extracting the Simplest Term

After we have plotted our expression, we are ready to *extract* the simplest term. The object is to *describe* the plotted squares with the *minimum* number of terms. Patterns of *four* squares can be described with a *one* variable term (A); *two* plotted squares are described by a *two* variable term (AB); and a *single* plotted square is described by a *three* variable term (ABC). Each of these is illustrated in Figure 7-43.

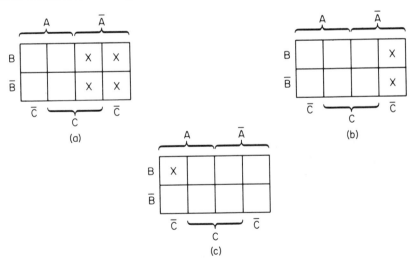

Fig. 7-43. Describe the Plots.

Example a shows *four* plots, and they are described by Ā. In example b, we have *two* plots, and they are described by $\overline{A}\overline{C}$. Example c has *only one* plot, and it is common to A, B, and $\bar{C}$. Our minimum expression for plot C is $AB\bar{C}$.

Since *more than one* term is generally plotted on a diagram, we usually describe the plot with more than one term. For example, the diagram in Figure 7-44 is a plot that requires *two* terms for a proper description.

We have two plots of two squares each which indicate that we have two terms of two variables each in our description of the plot. The plotted squares in Figure 7-44 are described by $A\bar{C} + \bar{A}B$.

Extract the simplified expression from Figure 7-45.

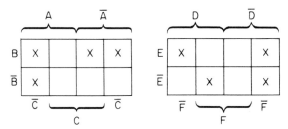

Fig. 7-44. Plot for $AB\bar{C}$ **Fig. 7-45.** Extract Simplified
$+ \bar{A}BC + A\bar{B}\bar{C} + \bar{A}B\bar{C}$.   Expression.

You should have obtained $\bar{D}\bar{F} + E\bar{F} + D\bar{E}F$.

We frequently encounter situations where the plot can be described by *more than one* simplified expression. Figure 7-46 is an example.

The three expressions that describe this plot are $A + \bar{A}$, $B + \bar{B}$, $C + \bar{C}$. Since they are equally *simple*, they are equally *correct*.

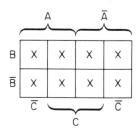

Fig. 7-46. Three Expressions Describe this Plot.

Plotting Four Variable Expressions

An expression with *four* variables requires a Veitch diagram of *16 squares*. The procedure for assignment of squares is the same as described previously. We start at the top and move counterclockwise, assigning half the squares to four variables and the other half to their complements. Each variable

Logic Design

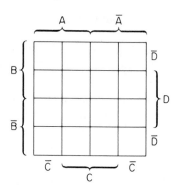

Fig. 7-47. Veitch Diagram for Four Variables.

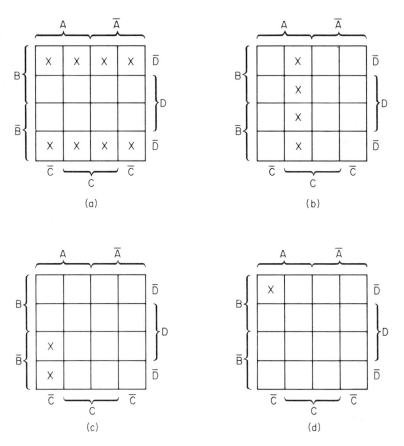

Fig. 7-48. Possible Patterns on 16 Square Veitch Diagrams.

overlaps all other variables and all complements except its own. This is illustrated in Figure 7-47.

The 16 square Veitch diagram *is not* limited to expressions of four variables. We can plot terms of one, two, three, and four variables on this diagram. *One* variable requires *eight* squares; *two* variables require *four* squares; *three* variables require *two* squares; and *four* variables require *only one* square. These four plots are illustrated in Figure 7-48.

Example A is a plot for the single variable $\bar{D}$; it uses *eight* squares. The plot for two variables (AC) is shown in example B; we use *four* squares for this plot. The three variable ($A\bar{B}\bar{C}$) plot is illustrated by example C; it requires *two* squares. Our four variable plot needs *only one* square. Example D is a plot for $ABC\bar{D}$.

Plot the expression $\bar{A}\bar{B}\bar{C} + \bar{A}B\bar{D} + \bar{B}\bar{C}D$, and compare with Figure 7-49.

Let's plot an expression containing terms of one, two, and three variables. Try this one: $\bar{A}CF + CFH + AF + \bar{A}\bar{H}$, and compare your results with Figure 7-50.

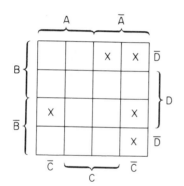

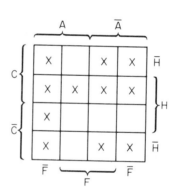

Fig. 7-49. Plot for $\bar{A}\bar{B}\bar{C} +$ $\bar{A}B\bar{D} + \bar{B}\bar{C}D$.

Fig. 7-50. Plot for $\bar{A}CF +$ $CFH + A\bar{F} + \bar{A}\bar{H}$.

Now let's plot an expression containing one, two, three, and four variables. Here is such a term: $A + \bar{A}\bar{B} + \bar{A}B\bar{C} + \bar{A}BC\bar{D}$. The plot is illustrated in Figure 7-51.

Figure 7-52 contains *two* plots. Each plot was made from an expression of *three* terms of *four* variables each. Study the plots and reconstruct the original expressions.

The *order* of your terms is not important, but here are the terms that you should have extracted:

a. $\bar{A}\bar{B}\bar{C}\bar{D} + ABC\bar{D} + \bar{A}\bar{B}CD$

b. $\bar{W}\bar{X}YZ + WX\bar{Y}\bar{Z} + WXYZ$

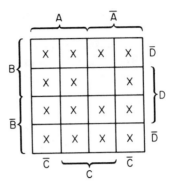

Fig. 7-51. Plot for $A + \bar{A}\bar{B} + \bar{A}B\bar{C} + \bar{A}BC\bar{D}$.

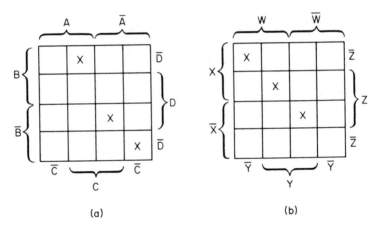

(a)   (b)

Fig. 7-52. Extract the Original Expressions.

Extracting Simplified Expressions

Patterns are formed by adjacent squares or squares on opposite ends of rows. The one, two, and three variable terms each form a pattern as illustrated in Figure 7-53.

Figure 7-54 contains *three* plots of Boolean expressions; extract the *simplest* expression to describe each of the plots.

The simplest expressions are:

a. $K + \bar{N}$

b. $QS + \bar{P}S$

c. $W\bar{X}$

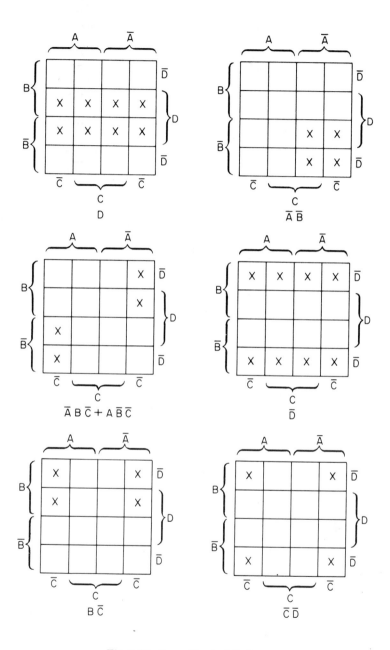

Fig. 7-53. Some Typical Patterns.

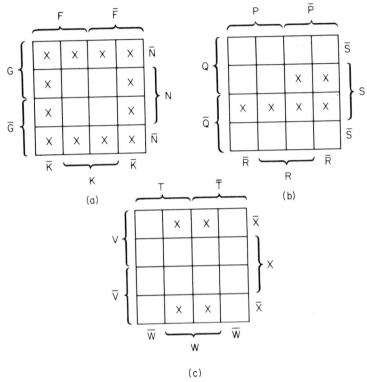

(a)

(b)

(c)

Fig. 7-54. Extract Simplest Expressions.

Extracting Complements of Plots

In many cases, we find that nearly all squares on our Veitch diagram have been plotted. It stands to reason that the *unplotted* squares represent the *complement* of the plotted squares. This complement can be very useful as a quick way of extracting the simplest expression. When the unplotted squares can be described by a *single term*, the complement method is the easiest, fastest, and most reliable method of obtaining the simplest expression. Figure 7-55 illustrates the method of extracting the complement of a plot.

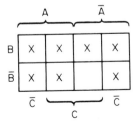

Fig. 7-55. Plot for A + B + C.

The object is to describe the square that is *not* plotted. What is it? It is obviously $\bar{A}\bar{B}C$. Now that we have the complement of the plot, what do we do with it? We *recomplement* by use of Boolean algebra. Place a vincula over the entire expression to indicate complement, then apply DeMorgan's theorem.

$$\bar{A}\bar{B}C$$

$$\overline{\bar{A}\bar{B}C}$$

$$A + B + \bar{C}$$

In this case, we traveled a full circle and came back to the expression that we plotted in the first place. This simply proves that the expression was already in its simplest form. Let's try this method on some expressions that are slightly more complex.

Plot this expression on a Veitch diagram: $\bar{B}C + \bar{A}D + \bar{C}\bar{D} + A\bar{B}\bar{C}D + ABC + \bar{A}B$. Check your results against Figure 7-56.

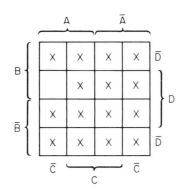

Fig. 7-56. Plot for $\bar{B}C + \bar{A}D + \bar{C}\bar{D}$
$+ A\bar{B}\bar{C}D + ABC + \bar{A}B$.

Our object is to find the simplest description of the plot. We'll follow these rules:

1. Extract the complement.
2. Complement the term extracted.
3. Apply DeMorgan's theorem.

The complement of the plot as represented by the unplotted square is $AB\bar{C}D$.

$AB\bar{C}D$	Extracted complement
$\overline{AB\bar{C}D}$	Add vincula
$\bar{A} + \bar{B} + C + \bar{D}$	DeMorgan's theorem

The expression, $\bar{A} + \bar{B} + C + \bar{D}$, is the simplest possible description of the plotted squares and the simplest version of the original expression.

Plotting Complements to Obtain Minterms

Some expressions are quite involved, and breaking them down to minterm form can be a lengthy process. Take this expression for instance: $\overline{A\bar{B}\bar{C} + \bar{A}\bar{B}\bar{D} + \bar{B}C + \bar{A}BC\bar{D}}$. After applying DeMorgan's theorm, we have $(\bar{A} + B + C)(A + B + D)(B + \bar{C})(A + \bar{B} + \bar{C} + D)$. This is still complex and we would need many applications of the distributive law before we could reach minterm form. What is the answer? We plot the *complement* of the original expression. What is the complement? It is the exact expression with the long vincula removed. Thus:

Original: $\overline{A\bar{B}\bar{C} + \bar{A}\bar{B}\bar{D} + \bar{B}C + \bar{A}BC\bar{D}}$

Complement: $A\bar{B}\bar{C} + \bar{A}\bar{B}\bar{D} + \bar{B}C + \bar{A}BC\bar{D}$

Plot the complement and compare your results against Figure 7-57.

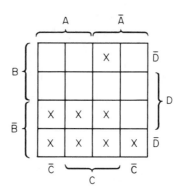

Fig. 7-57. Plot for $A\bar{B}\bar{C} + \bar{A}\bar{B}\bar{D}$ $+ \bar{B}C + \bar{A}BC\bar{D}$.

We have plotted the complement of our original expression. If we now extract the *complement* of the plot, we will have the original expression in simplified form. To do so, we describe the *unplotted* squares. In this case, we have: $AB + BD + B\bar{C} + \overline{AC}D$.

Complementing the AND Function

Perhaps you have noticed that all of our plotted expressions have been of the *overall* OR function. That is, if we converted them to logic diagrams, the final stage in each diagram would be an OR gate. Obviously we will encounter *overall* AND functions in expressions that need simplifying. Fortunately there is an easy way to do this. We will plot the *complement* and read the unplotted squares.

This is the pattern of an overall OR function:

$$\bar{B}C + \bar{A}\bar{B} + AB\bar{C}$$

This is the pattern of an overall AND function:

$$(J + \bar{K} + \bar{L})(H + \bar{J})(\bar{H} + K + L)$$

(The OR function is sometimes called a Sum of the Products, and the AND function then becomes the Product of the Sums. This is reverting the meaning of our signs of operation to standard mathematics, which is not a wise procedure).

Taking the complement of an OR function changes it to an AND function and vice versa.

We have this expression which we wish to simplify:

$$(\bar{A} + B + C)(\bar{B} + \bar{C})(A + \bar{B})(\bar{A} + \bar{B} + \bar{C})$$

Procedure:

1. Complement.
2. Apply DeMorgan's theorem.
3. Plot the result.
4. Describe the unplotted squares.

The indicated complement of $(\bar{A} + B + C)(\bar{B} + \bar{C})(A + \bar{B})(\bar{A} + \bar{B} + \bar{C})$ is $\overline{(\bar{A} + B + C)(\bar{B} + \bar{C})(A + \bar{B})(\bar{A} + \bar{B} + \bar{C})}$. After we apply DeMorgan's theorem, we have $A\bar{B}\bar{C} + BC + \bar{A}B + ABC$. This is in minterm form, and can be plotted as shown in Figure 7-58.

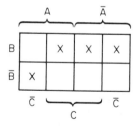

Fig. 7-58. Plot for $A\bar{B}\bar{C}$
$+ BC + \bar{A}B + ABC$.

The plotted squares in Figure 7-58 actually represent the *complement* of our *original* expression. So, we will describe the unplotted squares. This produces $\bar{B}C + \overline{AB} + AB\bar{C}$ which is the *simplest* form of our *original* expression.

Use the same procedure on the expression $(E + \bar{F})(\bar{E}G + \bar{F}\bar{H})(F\bar{G}H + \bar{E})$. It converts to $\bar{E}F + (E + \bar{G})(F + H) + (\bar{F} + G + \bar{H})E$. When we plot this we have the diagram in Figure 7-59.

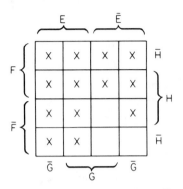

Fig. 7-59. Plot for $\bar{E}F + (E + \bar{G})$
$(F + H) + (\bar{F} + G + \bar{H})E$

When we describe the unplotted squares, we have the simplified version of our original expression which is $\bar{E}\bar{F}\bar{H} + \bar{E}FG$.

ANDing Veitch Diagrams

Sometimes we encounter expressions which are extremely difficult. Most of these become very simple when we *break* the expression into several *parts*. Consider this expression: $(ABC + A\bar{B}\bar{C})(A\bar{C} + \bar{A}\bar{B}C + ABC)(BC + \bar{A}B\bar{C})$. Considering the whole expression, it would be a long process to convert this to minterm form. Now, consider each parentheses *enclosed portion* as a separate expression. We now have these three parts: $ABC + A\bar{B}\bar{C}$ $A\bar{C} + \bar{A}\bar{B}C + ABC$ $BC + \bar{A}B\bar{C}$. Notice that each section is already in minterm form. Instead of ANDing the three sections together, we can accomplish the same thing by plotting each section on a separate diagram and ANDing the diagrams. The first plot is illustrated in Figure 7-60.

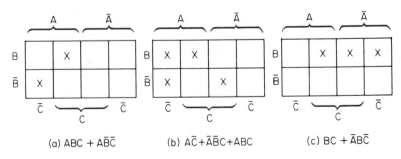

(a) $ABC + A\bar{B}\bar{C}$ (b) $A\bar{C} + \bar{A}\bar{B}C + ABC$ (c) $BC + \bar{A}B\bar{C}$

Fig. 7-60. A Plot for Each Section.

Each section of the expression is now plotted on a separate diagram. We could extract a simplified version of each section, but it is easier to obtain a simplified version of the *entire* original expression. We do this by ANDing the three diagrams. This is done by transferring the common plots to a fourth Veitch diagram. Consider the three diagrams in Figure 7-60. Isolate the plots that appear on *all* three diagrams. Only one square is *common;* a plot of this common plot is shown in Figure 7-61.

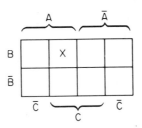

Fig. 7-61. The Common Plot.

When we describe this *common* plot, we have the *simplest* version of the *original* expression. In this case, it is ABC. So, instead of ANDing an expression, we plotted the sections of the expression, then ANDed the diagrams.

Review Exercises

1. What are the variables in computer circuits?

2. What constitutes a Boolean expression?

3. Illustrate four ways to indicate the AND function in a Boolean expression.

4. Name four signs that may express both grouping and function.

5. Name the three primary functions involved in Boolean expressions.

6. What type of switching circuit can be compared to these gates:
   (a) OR?
   (b) AND?

7. The inverter is frequently combined with another logic function. How is this indicated on a logic diagram?

8. What happens to our operation signs (connectives) when they are complemented?

9. Draw a logic symbol to describe each of these Boolean expressions:
   (a) $\bar{A}\bar{B}$
   (b) $\bar{A} + B + \bar{C}$
   (c) $\overline{A + B}$

10. What is the significant difference between $\overline{AB}$ and $\overline{A}\overline{B}$?

11. Draw a logic diagram showing four orders of logic.

12. What order of logic is described by each of these expressions?
(a) $[F(G + H) + I]J$
(b) $A + B + C$
(c) $(\overline{A} + B) + \overline{C} + D$
(d) $\overline{A + B}$
(e) $A(B + C) + D$

13. Write the proper Boolean expression for points 1, 2, 3, and 4 of the diagram in Figure 7-62.

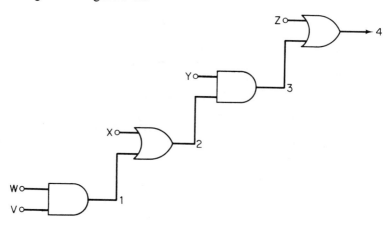

Fig. 7-62. Deriving Output Expressions.

14. Draw the logic for $\overline{F}\overline{G} + \overline{D}\overline{E}G$ and label outputs at each level.

15. (a) What order of logic is represented by Figure 7-63?
(b) What is the expression at point 1?

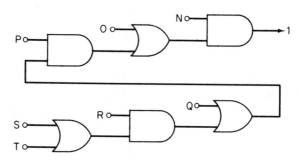

Fig. 7-63. Derive Output.

16. (a) Draw the logic for $[\overline{A\overline{B} + (C + D)}](\overline{E + \overline{F}})$.
(b) Label the output of each gate.

17. Compose the output expression for points 1, 2, 3, 4, 5, 6, and 7 for the logic in Figure 7-64.

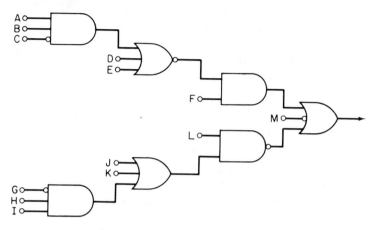

Fig. 7-64. Compose All Seven Outputs.

18. What is the advantage of simplifying Boolean expressions before designing the logic?

19. Which Boolean law is represented by each of these expressions:
(a) $A + AB = A$
(b) $AA = A$
(c) $\overline{\overline{A}} = \overline{A}$
(d) $\overline{\overline{A}} = A$
(e) $X = Y$
(f) $A\overline{A} = 0$
(g) $A \times 0 = 0$
(h) $\overline{A + B} = \overline{A}\overline{B}$
(i) $\overline{AB} = \overline{A} + \overline{B}$
(j) $A + BC = (A + B)(A + C)$
(k) $A(BC) = ABC$
(l) $A + 1 = 1$
(m) $X\overline{X} = 0$
(n) $A + \overline{A} = 1$
(o) $A(A + B) = A$
(p) $A \times 1 = A$
(q) $A + 0 = A$
(r) $A + (B + C) = A + B + C$
(s) $A(B + C) = AB + AC$

(t) AB = BA

(u) A + B = B + A

20. Write expressions to illustrate the two common identities which were derived from the basic Boolean laws.

21. Use Boolean laws to simplify these expressions:

(a) $\overline{(E + \overline{E})(\overline{FG} + \overline{G} + \overline{F})}$

(b) $(\overline{KL + M})(\overline{K} + \overline{L})\overline{M}$

(c) $[A + \overline{B}C + \overline{A}(B + \overline{C})]D$

(d) $AC + A\overline{C}$

(e) $(R + S)(T + V) + S(T + V)$

(f) $[\overline{A}(\overline{D} + B + D) + \overline{G} + BCD][A + G + \overline{A}\overline{D}EF]$

(g) $FG + EB\overline{B} + \overline{\overline{GF}}$

(h) $(B + C)(\overline{D} + B)(\overline{C} + B)$

22. (a) Derive the output at point 1.

(b) Simplify the output.

(c) Draw the logic for the simplified output.

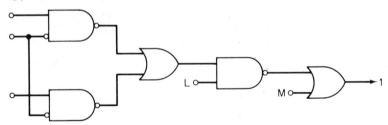

Fig. 7-65. Determine Simplified Output.

23. Define the expression "minterm"

24. What is a minterm type term?

25. Write all the minterms for the variables R, S, and T, without changing their order.

26. Which of these expressions are minterms for the variables A, B, C, D, and E?

(a) $\overline{A}\overline{B}\overline{C}\overline{D}\overline{E}$

(b) EDCBA

(c) BD + ACE

(d) $\overline{B}DCEA$

(e) A + B + C + D + E

27. Which of these expressions are minterm type terms for the variables J, K, L, and M?

(a) $JK\overline{L}$

(b) JK + LM

(c) JKLM

(d) $L\overline{M}$

(e) K

28. What is meant by a minterm expression?

29. Convert these expressions to minterm form:

(a) $\overline{T + V} + RS$

(b) $A + \overline{B} + \overline{CD} + \overline{A}\overline{B}C$

(c) $\overline{\overline{AB}} + CD + \overline{\overline{E}}B + \overline{BCD} + \overline{\overline{\overline{E}}}$

(d) $(DE + F)E + R(S + \overline{R})$

30. What are the possible number of minterms with:

(a) Three variables?

(b) Four variables?

(c) Five variables?

31. How many squares are required on a Veitch diagram?

32. How many squares would be required to plot each of these expressions?

(a) $\overline{J}\overline{K}L + \overline{M}N + M\overline{L}K$

(b) $A + DE + E\overline{A} + \overline{D}E\overline{A} + \overline{ED}$

(c) $V\overline{T} + X + \overline{Y} + TV\overline{X} + \overline{Y}V\overline{T} + \overline{X}$

33. Draw a Veitch diagram and plot $\overline{Z} + \overline{X} + \overline{Y}$.

34. Draw a Veitch diagram and plot $W\overline{X}\overline{Y} + \overline{W}XY + \overline{Y}WX + \overline{X}\overline{Y}\overline{W}$.

35. (a) What expression is plotted in Figure 7-66?

    (b) What is the simplest form of this expression?

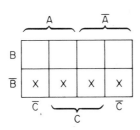

Fig. 7-66. What is Plotted?

36. (a) What expression is plotted in Figure 7-67?

    (b) What is the simplest version of this expression?

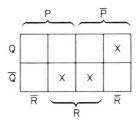

Fig. 7-67. Determine the Plot.

37. (a) Reduce this expression to minterm form

$$JK\bar{L} + \bar{J}K\bar{L} + \overline{J + \bar{K} + \bar{L}} + \overline{J + K + L} + L(\bar{J} + K)$$

   (b) Plot the new expression.
   (c) Extract the simplified expression.

38. Draw a Veitch diagram and plot $\bar{W} + X + Z$.

39. What expression is plotted in Figure 7-68?

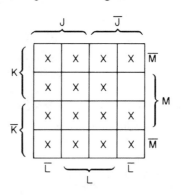

Fig. 7-68. Extract this Plot.

40. extract the simplied version of the expressions plotted in Figure 7-69.

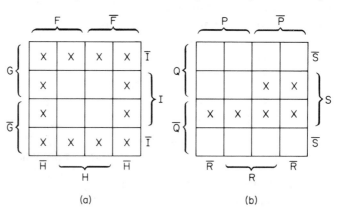

(a)　　　　　　　　　　　(b)

Fig. 7-69. Extract Simplified Expressions.

41. (a) Write the output expression for the diagram in Figure 7-70.
   (b) Convert the expression to minterm form.
   (c) Plot the expressions.
   (d) Extract the simplified expression.
   (e) Draw the simplified logic equivalent.

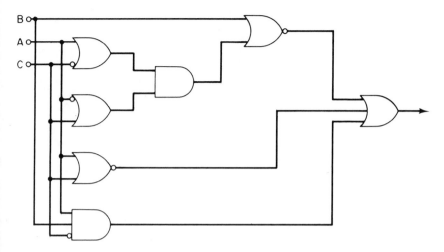

Fig. 7-70. Simplify.

42. What is the simplest form of:
    (a) $\overline{\overline{J}KL\overline{M} + \overline{K}L + J\overline{L}M + K + J\overline{K}}$?
    (b) $\overline{\overline{R}S + R\overline{S}\overline{T} + RT + RS\overline{T}}$?
    (c) $(\overline{Q}R + S\overline{T})(\overline{Q} + \overline{S})(R + \overline{S} + T)(RS + \overline{T})$?

43. Simplify this expression by ANDing Veitch diagrams: $(QR + \overline{Q}\overline{R}\overline{S}\overline{T})$ $(QR\overline{S} + \overline{R}\overline{S}\overline{T})(Q\overline{R} + \overline{S}T + \overline{Q}T + QR\overline{S})$ (Show all diagrams and the final expression).

chapter eight

functional
components

For our purpose, a functional component is a group of computer circuits that have been designed and constructed to perform a *specific function*. For instance, the computer clock is a functional component. It performs one task continuously: pulse production. We have several other functional components such as registers, counters, detectors, and converters. Each functional component performs one task over and over, and does it the same way every time. Numberwise, a computer may contain *thousands* of functional components; yet, there are relatively *few* distinct functions. This means that we have many repetitions of the same basic components. In this chapter we will examine the logic diagrams and operation of the most common functional components.

8-1 REGISTERS

A register is a device used for *storage* of a computer word of information. This may be a *temporary* storage for a few ns or a *permanent* storage that will remain until deliberately destroyed. We may group all registers into *two* categories: storage registers and shift registers.

Storage Registers

We have storage registers throughout the computer to store groups of binary bits. Any of several types of storage devices may be used to construct these registers. Some devices in use are flip flops, flip flop latches, ferrie cores, thin film spots, and even monostable (single shot) multivibrators. These registers *receive* the information on command, *preserve* it without loss of signal strength or alteration of configuration, and *deliver* the information when it is needed.

Particular binary bits are generally identifiable by their location in the register. When the computer operation so dictates, the register enables sampling of the bits either one at a time or as an entire group (word). The *storage area* of the central memory is composed of several thousand storage registers, and each register has a specific *address* and holds a *complete* computer word. As bits move about inside the machine they are relayed from one register to another.

Most types of storage devices have the capability of releasing a *copy* of the information they contain *without disturbing* the condition of the device itself. Thus, information can be transferred from one register to another and there will be a *copy* of the information in *both* registers. Devices of this type retain their information until we deliberately erase it. We refer to this erasing action as *clearing* the register. Ferrite cores and thin film spots are exceptions to this rule. When information is transferred from these devices the contents are erased.

In many cases, a *level* or a *pulse* on a line constitutes a binary one, and the *absence* of such a level or pulse constitutes a zero. This enables us to accomplish a transfer by activating only the lines that contain a one. Figure 8-1 illustrates this method of transfer.

In this drawing, we have two flip flop storage registers designated as A and B. Each register has four flip flops, and each flip flop holds one binary bit which is either a one or a zero. The clear pulse is applied directly to the clear input of all flip flops in the B register. When this pulse is applied, flip flops E, F, G, and H are all placed in the zero state. Assuming positive logic, the *zero state* gives us a *low* from the *one* side output and a *high* from the *zero* side. A flip flop in the zero state (cleared) contains a binary zero.

In the A register, we have completely ignored the zero side of the flip flops, but we have an AND gate between the one side output of each A flip flop and the set input of the corresponding B flip flop. Signal CP is the periodically occuring *clock* pulse. Signal TAB means *transfer* the contents of *A register* to *B register*. CP is a chain of sharp trigger pulses, TAB is a level of a

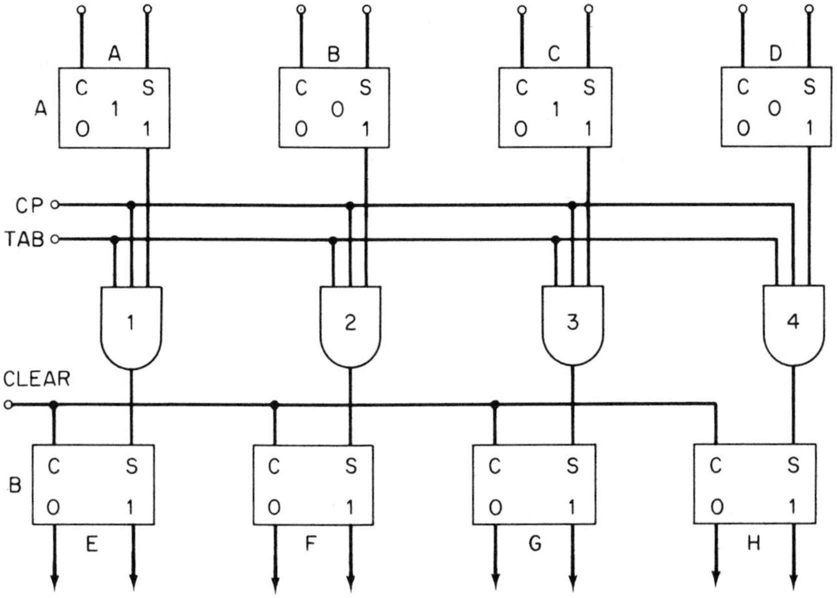

Fig. 8-1. Parallel, Ones Side Transfer.

duration sufficient to guarantee coincidence with at least one clock pulse, and the clear pulse is a gated pulse to be applied on command.

Assuming that flip flops A, B, C, and D contain binary information as indicated on the drawing, the transfer is accomplished as follows:

1. The clear pulse erases the contents of the B register by placing flip flops E, F, G, and H in the zero state.

2. Flip flops A and C contain ones and the one sides of these flip flops hold a conditioning level on one leg of AND gates 1 and 3.

3. Signal TAB arrives, and for a short time holds a *conditioning* level on one leg of all AND gates.

4. The next clock pulse, very briefly, *conditions* the third leg of all AND gates.

5. AND gates 1 and 3 are *activited* and conduct for the duration of the clock pulse, and apply a high to the set input of flip flops E and G.

6. Flip flops E and G will contain ones and flip flops F and H still contain zeros. We now have the same binary configuration in both registers.

If we were designing this type of logic, we would consider each flip flop individually and write a Boolean formula in terms of the *activated outputs* of each gate. For AND gate 1 the Boolean expression is (FF-A)(TAB)(CP). With the exception of the flip flop designation, we have the same expression for each of the AND gates. In this case, flip flop B contains a zero. The input to AND gate 2 is $(\overline{\text{FF-B}})(\text{TAB})(\text{CP})$. This leaves the AND gate *deactivated;* no transfer occurs. The same is true of flip flop D. However, the corresponding flip flops in the B register already contained zeros, so no transfer was necessary.

It is customary to name registers according to letters of the *alphabet* as we have done here, but we also find more definitive names for registers that serve distinct functions. Some of these are index register, address register, and memory information register. We identify specific bits in a register by *numbering the bits* in a word, from most significant to least significant bit. The storage devices in each register are then numbered in the same fashion.

Going back to Figure 8-1, the four bit register is representative; it can be any size. Reading from left to right, the A register contains 1010 with our MSB on the left. The bits are identified from left to right as 1, 2, 3, and 4. The flip flops in the A register are then identified from left to right as A1, A2, A3, and A4. The most common way of showing outputs from a flip flop as being a one is to simply show the number of that flip flop. In Figure 8-1, the one output of flip flops A and C can be shown by A1 and A3. Conversely, the zero output is designated by NOTting the position number. Thus flip flops B and D contain zeros, and we can show this by $\overline{A2}$ and $\overline{A4}$.

It is not difficult to arrange logic to *eliminate* the need of *clearing* the receiving register before each transfer. Such a transfer is known as *forced feeding* or forced loading of a register. The logic is illustrated in Figure 8-2.

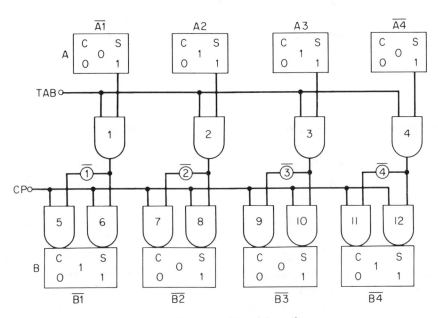

Fig. 8-2. Forced Feed Transfer.

With this arrangement, when the signal TAB occurs, the contents of the A register are *forced* into the B register, regardless of what the previous contents of B might have been. Let's consider the bits one at a time.

$\overline{A1}$ *deconditions* AND 1, providing a low output. This low is *inverted* through inverter 1 and becomes a *high* to AND 5. The clock pulse *satisfies* the other leg of AND 5 and applies a *high* to the clear side of flip flop B1. B1 was set (contained a one), but this input *clears* it (forces it to the zero state).

A2 *conditions* one leg of AND 2, and signal TAB conditions the *other leg*. AND 2 has a *high* out which conditions one leg of AND 8. The next clock pulse *activates* AND 8 and a *high* is applied to the set input of B2. B2 was clear but is now *forced* to the one state.

A3 and TAB *activate* AND 3. The high from AND 3 and CP activate AND 10 to set B-3.

$\overline{A4}$ *disables* AND 4 to produce a low output. Inverter 4 *changes* this low to a high which acts with CP to *activate* AND 11. This produces a high which *clears* B4.

The B register contained 1001 before the transfer (it could have been any configuration). These contents have been *forced out* and replaced by the contents of the transferring register, in this case, 0110.

Shift Registers

A shift register is a *storage* register with shifting *capabilities*. We can design such a register to perform any shifting operation we desire. Some examples are left, right, one bit at a time and all bits at once. A line may be connected from MSB to LSB and perform a *circular* or *end around* shift. Shift registers are used extensively in data manipulation areas like the arithmetic section.

When we use the shift register for simple storage, it can *receive* information in *either* serial or parallel form, and the information may be *extracted* in *either* serial or parallel form. This flexibility is extremely useful in many places in the computer system. For instance, parallel transfer is preferable to, from, and within the central computer because it is faster. On the other hand, some of the input and output data may be carried on telephone lines. This makes the serial transfer the only feasible way of handling data at the junction of the telephone line. A simple shifting action is illustrated in Figure 8-3.

This drawing illustrates the *right* shift as well as a *serial* output. We sometimes call this a *broadside* shift because all bits move at the same time. The three bit register is representative; a thirty bit or a sixty bit register works the same way.

Assume that we have A1, $\overline{A2}$, and A3 (101). Signal SP is our *shift* pulse. Notice that this signal enables one leg of all five AND gates. Flip flop A1 is *set*, which provides signal A1 to AND 5. Flip flop A2 is *clear*, and this $\overline{A2}$ disables AND 3, but a high from the zero side conditions *one* leg of AND 4.

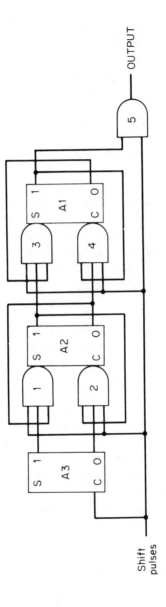

Fig. 8-3. Shift Right.

A1 is fed back to condition a *second* leg on AND 4. The high from the zero side of A2 is fed back to condition one leg of AND 3. A3 is *set*, and this provides a high to condition a *second* leg of AND 3. AND gates 1, 4, and 5 need only signal SP to activate them. When SP arrives, the one from A1 transfers out through AND 5; the zero from A2 transfers through AND 4 into A1; the one from A3 transfers through AND 1 into A2; and A3 is *cleared* to zero. Our configuration is now $\overline{A1}$, A2, and $\overline{A3}$, and we have transferred our most significant bit.

The next SP finds AND 5 *disabled* by $\overline{A1}$, and there is a low output from this AND gate. This condition serves in lieu of a zero transfer. The same SP moves the one from A2 into A1, the zero from A3 into A2, and leaves A3 in a *clear* state. We now have A1, $\overline{A2}$, and $\overline{A3}$.

The third SP *gates* A1 through AND 5. It also transfers the zero from A2 into A1. This leaves all flip flops clear, and our 101 has been shifted out (transferred) in serial form (one bit at a time).

Gating may be provided to *control what* data will be received, *when* it will be shifted, and *when* it will be transferred. This is illustrated in Figure 8-4.

Let's assume that all flip flops are *clear* with data in and data out lines connected to *telephone* lines. Although a flip flop may be designed to respond to either a positive transition or a negative transition of the signal, the general rule is a triggering on *negative transitions*. We will follow this general pattern unless otherwise specified. Data in, shifting, and data out are governed by the signals gate in (GI) and gate out (GO). Figure 8-5 is a timing chart which illustrates the actions in this register.

A *DATA IN* pulse arrives simultaneously with our first clock pulse. This conditions *two* legs of AND 2, but the third leg is deconditioned because both GI and GO are absent. All three flip flops remain in the *clear* state from CP1 to CP2. The same conditions repeat with CP2. GI rises at CP2, but the clock pulse is gone before it reaches the proper level. All flip flops are still *clear* at CP3.

At CP3, we have signal GI and DATA IN. DATA IN is changing from a one to a zero, but its change is *slow* compared to the clock pulse. On the negative transiton of CP3, AND 2 activates and sets A1. From CP3 to CP4, we have $\overline{\text{DATA IN}}$, A1, $\overline{A2}$, $\overline{A3}$, GI, and $\overline{GO}$.

At CP4, (GI)(CP4)(A1) *satisfy* AND 4 and set A2. At this same time, $\overline{\text{DATA IN}}$ is *inverted* to become DATA IN to AND 1. The signals (DATA IN) (CP4)(GI) *activate* AND 1 and A1 is *cleared* to $\overline{A1}$. During the clock pulse, DATA IN was changing from a zero to a one, but the action is complete *before* this change can be accomplished. From CP4 to CP5, we have DATA IN, $\overline{A1}$, A2, $\overline{A3}$, GI, and $\overline{GO}$. We have *shifted* the one bit from A1 to A2 and shifted a zero bit from the input line to A1.

At CP5, (GI)(CP5)(A2) *activate* AND 6 and set A3; (GI)(CP5)$\overline{A1}$ activate AND 3 and clear A2; and (DATA IN) (GI)(CP5) activate AND 2 and set A1.

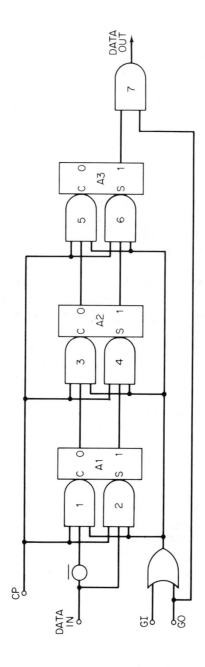

Fig. 8-4. Serial In, Serial Out Shift Register.

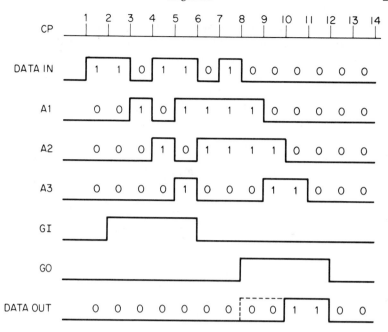

Fig. 8-5. Timing Chart for Figure 8-4.

At this time, we shifted a one from A2 to A3, a zero from A1 to A2, and a one from the input line to A1. From CP5 to CP6, we have DATA IN, A1, $\overline{\text{A2}}$, A3, GI, and $\overline{\text{GO}}$.

At CP6, DATA IN and GI start toward zero, but triggering will occur *before* the transition can be made. Signal A3 conditions one leg of AND 7, but there is *no output* because the GO signal has not arrived. Since the *absence* of a one constitutes a zero, our DATA OUT has been a continuous zero. As long as we have $\overline{\text{GO}}$, we will have $\overline{\text{DATA OUT}}$.

Signals (GI)(CP6)($\overline{\text{A2}}$) activate AND 5 and clear A3; (GI)(CP6)(A1) activate AND 4 and set A2; and (DATA IN) (CP6)(GI) activate AND 2 and set A1. From CP6 to CP7, we have $\overline{\text{DATA IN}}$, A1, A2, $\overline{\text{A3}}$, $\overline{\text{GI}}$, and $\overline{\text{GO}}$.

At CP7, we have $\overline{\text{GI}}$ and $\overline{\text{GO}}$. *None* of the AND gates will be activated at this time. A3 remains clear, A2 and A1 remain set, and the zero input ($\overline{\text{DATA IN}}$) *is not* transferred in.

At CP8, we have *almost* a repeat of CP7 conditions. GO starts up but *not in time* for gating. All flip flops remain unchanged, and the one bit on the data input line is lost. From CP8 to CP9, we have $\overline{\text{DATA IN}}$, A1, A2, $\overline{\text{A3}}$, $\overline{\text{GI}}$, and GO.

At CP9, signal (GO)(CP9)(A2) activate AND 6 and set A3; (GO)(CP9)(A1) activate AND 4 to the set side of A2 (A2 remains set); and $\overline{\text{DATA IN}}$ is *inverted* to DATA IN. (DATA IN)(CP9(GO) activate AND 1 and clear A1.

At this time, we shifted a one from A2 to A3, a one from A1 to A2, and a zero from the input line to A1. AND 7 *has not* yet been activated, therefore we still have zero for data out. From CP9 to CP10, we have $\overline{\text{DATA IN}}$, $\overline{\text{A1}}$, A2, A3, $\overline{\text{GI}}$, and GO.

At CP10, signals (GO)(CP10)(A3) satisfy AND 7 and gate out our *first* one bit. At the same time, (GO)(CP10) (A2) activate AND 6; (GO)(CP10)($\overline{\text{A1}}$) activate AND 3 and clear A2; and [DATA IN (INVERTED $\overline{\text{DATA IN}}$)] (CP10)(GO) activate AND 1. A3 remains set, and A1 remains clear. At this time, we shifted a one from A3 to the output line, a one from A2 to A3, a zero from A1 to A2, and a zero from the input line to A1. From CP10 to CP11, we have $\overline{\text{DATA IN}}$, $\overline{\text{A1}}$, $\overline{\text{A2}}$, A3, $\overline{\text{GI}}$, and GO.

At CP11, (GO)(CP11)(A3) activate AND 7 and transfer our *second* one bit. At the same time, (GO)(CP11)($\overline{\text{A2}}$) activate AND 5 and clear A3. From CP11 to CP12, all flip flops are clear giving us $\overline{\text{DATA IN}}$, $\overline{\text{A1}}$, $\overline{\text{A2}}$, $\overline{\text{A3}}$, $\overline{\text{GI}}$, and GO.

At CP12 the gate out drops to place all our signals in the *not* state. Chances are that the GO signal is intended to select discrete samples of the output signal at specific times. If so, in this case, we sampled (left to right) 0011. The signal GI and GO work together to select the times that incoming data will be shifted in. Notice that A1 *does not* change with either CP7 or CP8 even though input data does change at these times.

Shifting Operations

When a computer performs multiplication or division the result contains *twice* as many binary bits as a normal computer word. Multiplication is generally performed by alternately adding and shifting. Division is accomplished by alternately adding the ones complement (subtraction) and shifting. To avoid loss of half the bits from these operations our computer must have a shift and store capability of *two computer words* in series. We generally accomplish this by connecting two shift registers in series. We call this a double shift operation. The register arrangement is illustrated in Figure 8-6.

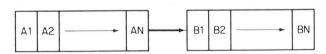

Fig. 8-6. Double Shift.

At the conclusion of a multiplication or division, both registers are filled, with A1 being the most significant bit (MSB) and AN being the least significant bit (LSB). Assuming that we have a 30 bit computer word, we have now solved a problem and generated a 60 bit result. Since the storage registers in memory *cannot* accommodate double length words, this result must be *reduced* to normal size before it can be stored. Actually reduction is a very simple operation. Bit B1 is sampled to see if it is a one or a zero. When B1 is a *one*, we *add one* to the LSB of the A register (BN). This is proper mathematical procedure for *rounding off* a number. When B1 is a one, the total contents of the B register must be at *least* $\frac{1}{2}$. So when we add one to the LSB of the A register, we overcompensate slightly for the 30 bits we are *discarding* from the B register. When we find B1 is a zero, the contents of the B register must be *less than* $\frac{1}{2}$. In this case, we simply take the A register contents without any change.

When only one register is involved in a shifting operation, we call it a *single* shift. When two registers are involved, it is a *double shift*. There are *four* variations to both the single and the double shift. These are:

Left

Right

End off

End around

A particular shifting action is composed of either *left* or *right* combined with either *end off* or *end around*. All of these combinations for the single shift are shown in Figure 8-7.

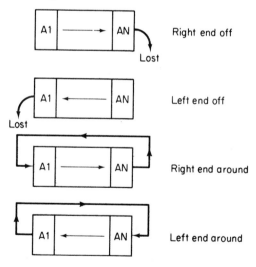

Fig. 8-7. Variations of Single Shift.

During the *left end off* shift, we discard bits from the MSB position of the register. In the *right end off* shift, we discard bits from the LSB position of the register. In either case, the number of bits *lost* is equal to the number of *shifts* performed in the operation.

During the *end around* shifts, we lose no bits. The *right* shift moves the LSB to the MSB position, and the *left* shift moves the MSB to the LSB position. The number of bits *exchanged* in this fashion is equal to the number of shifts performed in the operation.

The double shift variations are the same. The only difference being the tandem hookup of the two resisters. These variations are shown in Figure 8-8.

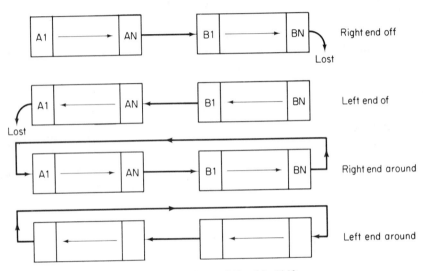

Fig. 8-8. Variations of Double Shift.

In most cases, our computer word contains one binary digit which identifies the number as *either* a positive or a negative quantity. We call this bit a *sign bit*. In most cases, the sign bit appears in the position which would otherwise be the MSB position of the register. This is illustrated in Figure 8-9.

S I G N	A1	A2		AN

Fig. 8-9. Position of the Sign Bit.

In most computers, we find that a *zero* sign bit designates the word as a *positive* quantity while a *one* tells us that the quantity is negative. Bringing the sign bit into our shifting operations gives us *two* more variations for both single and double shifts. When the sign bit is shifted with the other bits, we call it a *logical* shift. When the sign bit *does not* shift, we call it an *arithmetic*

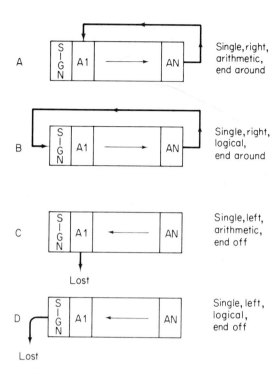

Fig. 8-10. Logical and Arithmetic Shifts.

shift. Examples of both logical and arithmetic shifts are illustrated in Figure 8-10.

Only a few of the variations are shown here. A and B can both become left shifts by reversing the arrows. C and D can each be changed to right shifts. All eight of these variations have counterparts in the double shift operation.

Many shifting operations are strictly *hardware* functions. The registers are designed to perform only one type of operation. Many other shifting operations can be *programmed.* The variations of the single and double shift are generally programmed functions. When we program a specific shift, we use an instruction to specify precisely what shift is desired. We specify single or double, left or right, logical or arithmetic, end off or end around, and the number of shifts to be performed. One shift moves all bits by one bit position.

8-2 COUNTERS

Counters form indispensable sections of *all* computer control elements. They count *addresses* for sequential access to memory locations. They count *words* that are transferred between memory and input/output devices. They count reiterative *steps* in a program. A counter can be designed to count either up or down, and some counters can do both.

Serial Up Counter

The basic configuration of the serial up counter is shown in Figure 8-11.

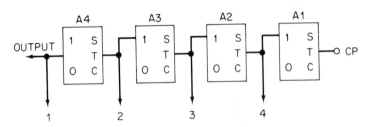

Fig. 8-11. Serial Up Counter.

We generally construct counters to respond to the *negative transition* of the input signal. This four stage counter counts clock pulses and provides an output from A4 on the eighth (decimal) clock pulse. The output is present between pulses 8 and 16, and pulse 17 starts the count over again. The timing is illustrated in Figure 8-12.

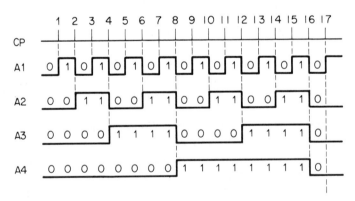

Fig. 8-12. Timing for Fig. 8-11.

The voltage levels at 1, 2, 3, and 4 are used to light indicators. When read in this order, these indicators show the *total count* in the counter at each clock pulse.

CP1 sets A1. CP2 clears A1 and the negative transition of A1 sets A2. CP3 sets A1. CP4 clears A1; the negative transition of A1 clears A2; and the negative transition of A2 sets A3. (Count is 0100). CP5 sets A1. CP6 clears A1 and sets A2. CP7 sets A1. CP8 clears A1, clears A2, clears A3, and sets A4. (Count is 1000). CP9 sets A1. CP10 clears A1 and sets A2. CP11 sets A1. CP12 clears A1, clears A2, and sets A3. (Count is 1100). CP13 sets A1. CP14 clears A1 and sets A2. CP15 sets A1. (Count is 1111). CP16 clears all four flip flops. CP17 starts the count over at 0001. The count at the termination of any clock pulse can be determined by reading the chart top to bottom at that time. For instance, the count in the counter between CP10 and CP11 is 0101.

Serial Down Counter

The only difference between the serial up and serial down counter is the fact that the down counter uses the zero outputs from the flip flops. The up counter *increases* its count to a *maximum* (all flip flops set), clears to zero and starts up again. The down counter *decreases* its count to a *minimum* (all flip flops clear) resets to a full count (all flip flops set) and starts down again. The serial down counter is illustrated in Figure 8-13.

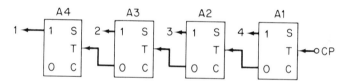

Fig. 8-13. Serial Down Counter.

We start with a count of 15_{10} (all flip flops set). The indicators 1, 2, 3, and 4 are all lit to show the binary configuration 1111. CP1 clears A1 and reduces the count to 1110. CP2 sets A1, and the negative transition of A1 (zero side), clears A2. The count is now 1101. The action continues in the same fashion until the count is 0000. This requires 15_{10} clock pulses. CP16 sets all flip flops for a count of 1111. CP17 reduces the count by one to start the action over again.

A counter of this type may be employed to keep track of the number of shifts being accomplished or the number of words being transferred. In this case, the desired number is *set* into the counter at the *start* of the operation. If we are shifting, each shift *reduces* the count by one, and when the count reaches zero, it signals the *end* of the shifting operation.

In high speed operations, serial counters are at a disadvantage. When a pulse initiates an action which changes the state of several flip flops, the first flip flop completes its action before the second starts; the action in the second is completed before the third begins. The total time involved, in each case, is the sum of the switching times of all flip flops that take part in that action. When our counter has many stages, the time interval can be considerable. The solution to this time problem is provided by parallel counters.

Parallel Up Counter

The parallel counter has two distinct features. The clock pulse (or pulse being counted) is applied to *each* stage, and the output of each flip flop is applied to *all* higher order flip flops. This arrangement allows any count to be inserted or altered in the time required for one flip flop to change states. A parallel up counter is illustrated in Figure 8-14.

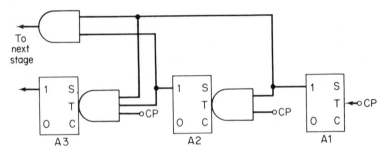

Fig. 8-14. Parallel Up Counter.

Starting with all flip flops *clear*, the first clock pulse sets A1. CP2 clears A1 and sets A2. CP3 sets A1. CP4 clears A1 and A2 and sets A3. CP5 sets A1. CP6 clears A1 and sets A2. CP7 sets A1, CP8 clears *all* three flip flops and sets the next higher order flip flop. This timing is shown in Figure 8-15.

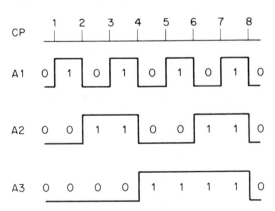

Fig. 8-15. Timing for Fig. 8-14.

Parallel Down Counter

The parallel down counter untilizes the *zero* side outputs of the flip flops. Otherwise it is identical to the up counter. This is illustrated in Figure 8-16.

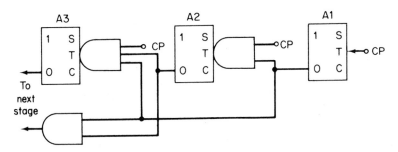

Fig. **8-16.** Parallel Down Counter.

Starting with all flip flops *set*, CP1 clears A1. CP2 sets A1 and clears A2. CP3 clears A1. CP4 sets A1 and A2 while clearing A3. CP5 clears A1. CP6 sets A1 and clears A2. CP7 clears A1. CP8 sets *all* three of these flip flops and *clears* the next higher order flip flop. Timing for this counter is shown in Figure 8-17.

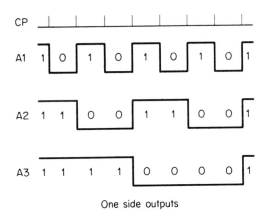

One side outputs

Fig. **8-17.** Timing for Fig. 8-16.

Count in a Counter

If we allow n to equal the number of *flip flops* in a counter, then two to the nth power (2^n) is the total number of *possible* conditions. Thus a four stage counter has 2^4 or 16 (decimal) possible conditions. This is 0000 through

1111. The maximum count, in this case, is 1111 (or 15_{10}). In any case, the maximum count is 2^n-1.

At times it is essential to be able to *determine* the count in a counter after a specified number of inputs. It is a simple procedure to start with any specified count, apply any number of pulses, and quickly determine the number of cycles as well as the count left in the counter.

For an up counter, we have:

$$CY+NC = IP+OC/2^n$$

Where: CY is the number of counter cycles, NC is the new count, IP is the number of input pulses, OC is the original count in the counter, 2^n is the maximum number of conditions in a cycle, and n is the number of flip flops in the counter.

When we solve a problem using this formula, we generally have a *whole* number and a *remainder*. The whole number is the number of *cycles*, and the remainder is the *new count* in the counter.

Problem: We have a five stage up counter which contains 10100_2. After application of 20_{10} pulses, what is the *new count*, and how many *cycles* occurred?

Solution: $CY+NC = IP+OC/2^n$
(All numbers $CY+NC = 20+20/32$
in decimal) $CY+NC = 40/32$
                        $CY+NC = 1$ with a remainder of 8
                        $CY = 1$
                        $NC = 8$

For a down counter, we use the same symbols with a slightly different arrangement.

$$CY + \text{remainder} = IP\text{-}OC/2^n$$
$$\text{The new count (NC)} = 2^n\text{-}R$$

Problem: We have a five stage down counter with a binary configuration of 10100. After application of 69_{10} pulses, what is the *new count*, and how many *cycles* occurred?

Solution: $CY+R = IP\text{-}OC/2^n$
(All numbers $CY+R = 69\text{-}20/32$
in decimal) $CY+R = 49/32 = 1$ and R of 17
                        $CY = 1$
                        $NC = 32 - 17 = 15$

Modulus Counters

Since each flip flop added to a counter *doubles* the maximum count, we must have a way to terminate the count some place between the normal flip

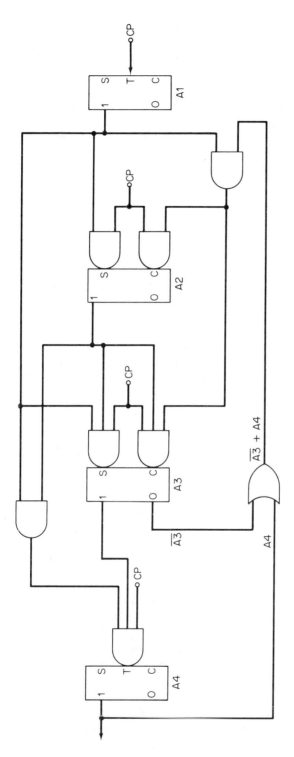

Fig. 8-18. Modulus 10 Counter.

flop outputs. For instance, we need to count to 10_{10} and then recycle the counter. A three stage counter can count to 7 and recycle while a four stage counter counts to 15_{10} and recycles. We must have a *four* stage counter, *but* we want to recycle on the count of 10_{10}. Such a counter is called a *modulus* counter and a sample is illustrated in Figure 8-18.

The object is to produce one output from A4 for every 10_{10} pulses received at A1. Starting with all flip flops *clear*, CP1 sets A1. CP2 clears A1 and sets A2. CP3 sets A1. CP4 clears A1 and A2 while setting A3. CP5 sets A1. CP6 clears A1 and sets A2. CP7 sets A1. CP8 clears A1 and sets A4. CP9 sets A1. CP10 clears *all* flip flops. The timing for this counter is shown in Figure 8-19.

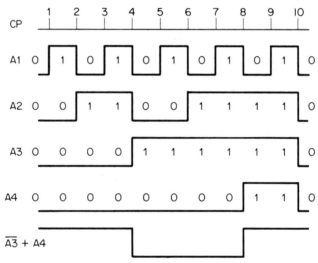

Fig. 8-19. Timing for Fig. 8-18.

We encounter *many* modulus counters in computers. The *most common* are modulus three, modulus 10, and modulus 14.

Ring Counters

In computer control, we allot a specified period of time for an operation to take place. This may include *removing* an instruction from memory and decoding it. It may include *obtaining data* from memory and operating on an instruction after it has been decoded. It may include *storing* a result in memory after an instruction has been performed. In some computers, all three of these operations may be performed as a *single* operation.

We call this period of time a *cycle*, and the cycle is divided into a specific number of timing intervals. The clock pulse is sufficient to trigger most actions, but we need a means of specifying when a cycle starts and when it ends. The *ring counter* is generally used for this purpose.

We construct the ring counter with one flip flop for each timing pulse (TP) required during a standard timing cycle. The flip flops are interconnected so that *only one* flip flop can be set at a given time. It counts the clock pulses from the start to the end of each cycle. This counter is illustrated in Figure 8-20.

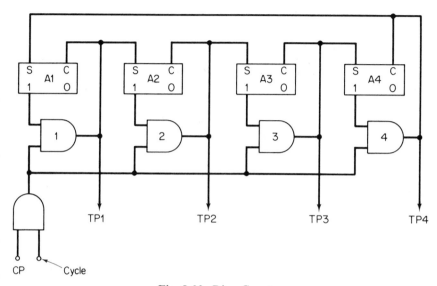

Fig. 8-20. Ring Counter.

This is an *abbreviated* counter. In actual practice, we would probably use twelve or more flip flops.

We stop each cycle with A1 set. This holds a conditioning level on one leg of AND 1 until we start the next cycle. The first clock pulse of each cycle activates AND 1. The output of AND 1 clears A1, sets A2, and at the same time, gives us the *first* timing pulse of the cycle. AND 2 is now conditioned, and CP2 passes through. The output of AND 2 clears A2, sets A3, and becomes TP2. This conditions AND 3. CP3 passes through AND 3 to clear A3, set A4, and become TP3. This conditions AND 4. CP4 passes through to clear A4, set A1, and become TP4. This *terminates* the cycle and leaves A1 set. Regardless of the number of stages used, the feedback to A1 will coincide with the final output timing pulse of the cycle.

$$TP1 = CYCLE \cdot CP \cdot A1$$
$$TP2 = CYCLE \cdot CP \cdot A2$$
$$TP3 = CYCLE \cdot CP \cdot A3$$
$$TP4 = CYCLE \cdot CP \cdot A4$$

8-3 DETECTORS

Many operations require our computer to *sample* the contents of registers or counters and to *initiate* a logical action based on the condition that is present at that instant. This sampling is performed by *detectors* which are sometimes called decoders.

AND Gate Detector

We may use AND gates to detect the *presence* of a specific number or any desired combination of numbers. The output of the final AND gate can be a trigger, or a level, which occurs only when all the specified conditions are present. A simple AND gate detector is illustrated in Figure 8-21.

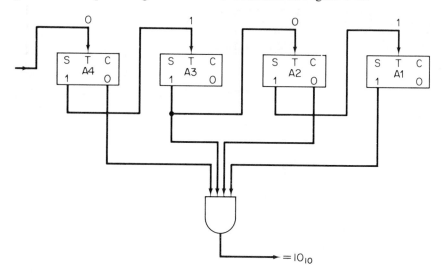

Fig. 8-21. AND Gate Detector.

This AND gate is connected to selected outputs from the flip flops of a counter. With the connections we have, the AND gate will *detect* a decimal count of 10. This is a binary configuration of 1010 as shown above the flip flops. An *output* from the AND gate signals the condition $A1.\overline{A2}.A3.\overline{A4}$. This is the *only* combination that will activate our AND gate.

Suppose that we wish to detect some number *other than* 10. Then we use a different Boolean formula. We construct the formula from the binary configuration which we wish to detect. To detect 1110 (14_{10}), our formula is $A1.A2.A3.\overline{A4}$. The inputs to the AND gate are connected from the *one* side of the *set* flip flops and the *zero* side of the *clear* flip flop. In this case, we use the one side of A1, A2, and A3 and the zero side of A4.

Combination AND-OR Detector

With an AND-OR combination, we can detect both the *presence* of a number and the *absence* of the same or any other number. One such combination is shown in Figure 8-22.

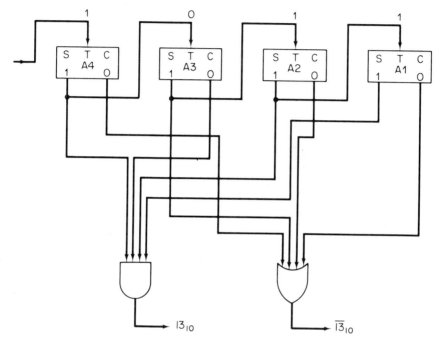

Fig. 8-22. AND-OR Detector.

Here we are detecting *both* the presence of 13_{10} and the absence of that number. The AND gate is activated *only* by the binary configuration 1101. The OR gate is activated by *any* configuration *except* 1101. The Boolean equation for the AND gate is $A1 \cdot A2 \cdot \overline{A3} \cdot A4$. The OR gate is activated by $\overline{A1} + \overline{A2} + A3 + \overline{A4}$.

AND/OR detectors are frequently called *diode* detectors because the AND and OR circuits are generally composed of solid state diodes.

Matrix Detector

It is frequently necessary to detect both the presence and absence of *all possible* binary configurations. Two areas requiring this type of detailed detection are *memory addressing* circuits and circuits which decode our *computer instructions*. Our detector networks are rather extensive in these cases, and they take on the physical appearance of mathematical matrices. We call such arrangememts *matrix* detectors. We must have four AND gates and four OR gates to detect all possible configurations in just two flip flops. This matrix is shown in Figure 8-23.

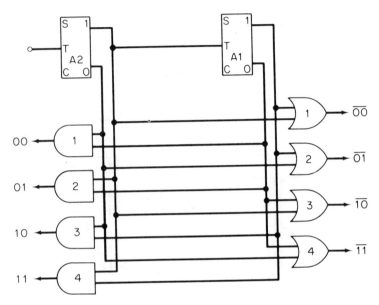

Fig. 8-23. AND-OR Matrix.

The AND gates detect the *four* possible true conditions. They have outputs *only* when the condition is present.

$00 = \overline{A1} \cdot \overline{A2}$ detected by AND 1

$01 = \overline{A1} \cdot A2$ detected by AND 2

$10 = A1 \cdot \overline{A2}$ detected by AND 3

$11 = A1 \cdot A2$ detected by AND 4

The OR gates detect the *four* possible NOT conditions. They have an output at *all times except* when the complement of the condition is present. The NOT detection can be confusing if we are not careful. Remember that the Boolean expression is in terms of the *true* output. When either A1 or A2 is high, OR 1 is activated, and the output verifies that the configuration is NOT 00. When the condition 00 is present, we have the *complement* of the activating condition, and OR 1 cuts off.

$\overline{00} = A1 + A2$ detected by OR 1

$\overline{01} = A1 + \overline{A2}$ detected by OR 2

$\overline{10} = \overline{A1} + A2$ detected by OR 3

$\overline{11} = \overline{A1} + \overline{A2}$ detected by OR 4

Matrices are generally shown in *schematic* rather than logic form. When we draw the schematic for Figure 8-23, it takes on the appearance of a matrix. This is shown in Figure 8-24.

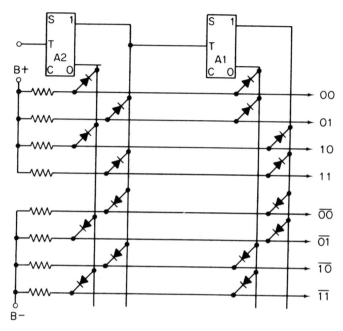

Fig. 8-24. Schematic of Fig. 8-23.

The *eight* diodes which are connected to the B+ voltage source form our *four* AND gates. One AND gate is composed of *two* diodes and a common resistor. The anodes go to B+, and the cathodes to the flip flop output. When *either* diode conducts the output is low. When we have a *high* on the cathodes of both diodes, they cut off, and the output line goes to B+. This constitutes an *activated* output.

The *eight* diodes which have their cathodes returned to B— form our *four* OR gates. When *either* diode of a given gate conducts, the voltage drop across the common resistor is high enough to constitute a *high* output (higher than B—). When *both* diodes cut off because of high potentials on the cathodes, the output line becomes B— which constitutes the NOT (or false) output.

Even when we detect *only* the true configuration, AND gates only, the diode matrix grows at the rate $n2^n$ where n is the number of variables. Detecting the *true* outputs of three flip flops requires 3×2^3, or 24, diodes. A detector for 12 flip flops requires $12 \times 12^{12} = 12 \times 4096 = 49,152$ diodes. This becomes a design problem because all diodes that we connect to a common line are in *parallel*. While the back current from one diode is negligible, the back current of several diodes in parallel becomes highly significant.

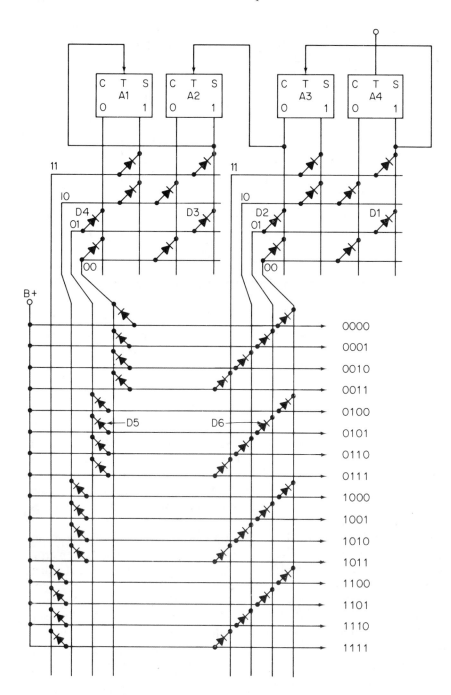

Fig. 8-25. Treed Diode Matrix.

We use two techniques to keep back current under control: add sneak path diodes in series and divide the matrix into several subsections.

Treed Matrix

Treed matrix is a term we apply to a matrix which uses *two* logic levels of AND gates. Again, the name shows little imagination. It is so called because of its branching appearance. In the treed arrangement, we divide the flip flops in *pairs* and combine the outputs of *a pair* through AND gates. Another section of AND gates *combine* the ANDed outputs from *all pairs*. The chief advantage of this arrangement is *conservation* of diodes, but it also helps to eliminate excessive back current. The double AND arrangement is illustrated in Figure 8-25.

Each *four digit* output is compiled through two levels of AND gates. The *numbered* diodes are used for detecting the configuration 0101. D_1 and D_2 are the input legs to the first AND gate. $01 = A4 \times \overline{A3}$. D_3 and D_4 are the legs for our second AND gate (still first level logic). In this case $01 = A2 \times \overline{A1}$. The outputs from the *two first level* AND gates feed into a *second level* AND gate which is composed of D_5 and D_6. The formula for cutting off all six diodes simultaneously is $\overline{A1} \times A2 \times \overline{A3} \times A4$. When this occurs, we have a high output on the 0101 line.

8-4 ENCODERS

An encoder function is the *converse* of the decoder or detector. We have seen that a detector translates a *binary* configuration into a *single* output signal. The encoder takes a *single* input and translates it into a binary *configuration*. One of the common computer input devices is a *keyboard* similar to that of a common typewriter. When you think of the alphanumeric keys on such a keyboard, the need for encoders becomes obvious. Figure 8-26 illustrates a simple method of *encoding* decimal numbers from zero through nine.

When we push down a key, it connects the *positive* voltage to appropriate legs on the OR gates. Pushing zero provides no input, and *all* OR gates have a low output. This produces the binary code of 0000. When we push the key for *number 5*, we activate A and C for *high* outputs, but B and D remain deactivated for *low* outputs. This produces the binary code 0101. Pushing keys 3, 4, and 8 in that order, produces this binary configuration: 0011 0100 1000.

8-5 CONVERTERS

In the central computer (memory, control, and arithmetic elements) our language is composed of *binary* digits in the form of standard amplitude pulses or voltage levels. When our central computer *exchanges* information

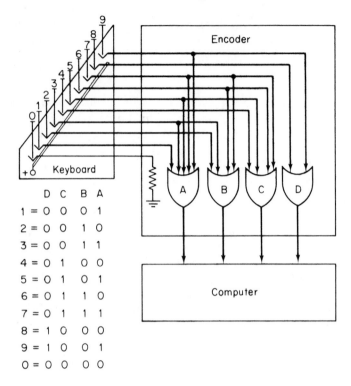

Fig. 8-26. Decimal to Binary Encoder.

with certain other devices, particularly the display element, the digital information is *not* always adequate. A certain binary configuration might indicate to the computer that a missile is off course, but these binary digits *cannot* draw a picture on the CRT to show how much it is off course. To do this, we need to *convert* the digital information into specific quantities of voltage and apply these to the CRT deflection plates. This is accomplished by a circuit network called a *digital to analog* converter.

Reversing our position and looking at the problem from the display end, we have the same problem in reverse. The picture is in a certain position on the scope because precise *values* of voltage are causing it to be painted there. When we *relay* this information to the central computer, we must *convert* these voltage levels to *digital* form. This task is accomplished by an *analog to digital* converter.

Digital to Analog

There are many variations of the digital to analog converter. Perhaps the *simplest* of these is the relay and resistor network. This type of converter is illustrated in Figure 8-27.

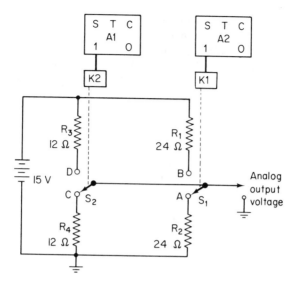

Fig. 8-27. Digital to Analog Converter.

Here we are converting *two* binary bits to *analog* voltage. The resistive network can be enlarged to convert *any* number of bits. K_1 and K_2 are relays. They are *deenergized* when our associated flip flops contain a *zero*, and *energized* when the flip flops contain *ones*. K_1 is magnetically connected to S_1 and K_2 to S_2. S_1 and S_2 are *shown* in the deenergized position. This means that both of our flip flops are *clear* and both relays are *deenergized*.

The switches are returned to ground through R_2 and R_4, and there is *no current* in the circuit. The analog voltage output is *zero*. We may gather from this that the output is zero volts when *all* binary bits are zeros.

When we *set* A2, K_1 *energizes* and moves the contacts of S_1 from A to B. We now have a *series circuit* consisting of the battery, R_1, and R_4. Our analog voltage is developed across R_4, and $E_{out} = 15\text{ V} \times 12/36 = 5\text{ V}$. So, a count of 01 *converts* to a voltage which is equivalent to 1/3 of the 15 V applied.

When we clear A2 and set A1, K_1 *deenergizes* and K_2 *energizes*. S_1 moves from B to A, and S_2 moves from C to D. We still have a series circuit, and now it is composed of the battery, R_3, and R_2. The output is now developed across R_2 which is 2/3 of the total resistance. For a count of 10, our analog voltage $= 2/3 \times 15\text{ V} = 10\text{ V}$.

When both bits are *ones*, A1 and A2 are *both* set. This *energizes* K_1, moving S_1 from A to B, and K_2, moving S_2 from C to D. Both *switches* are now returned to the positive side of the battery. The circuit is *incomplete*, but the *output* is effectively taken from the positive battery terminal with respect to ground. E_{out} becomes the same as E_a. We may conclude that when *all* binary bits are *ones*, the analog output voltage will be the *same as E_a*.

The circuit of Figure 8-27 may be *enlarged* to handle inputs from several flip flops. The flip flops would be added on the right as A3, A4, etc. As we moved toward the least significant bit. Another branch and another switch would be added for *each* flip flop. The resistor values in each added branch must be exactly *twice* the resistance of the previous branch. This is illustrated in Figure 8-28.

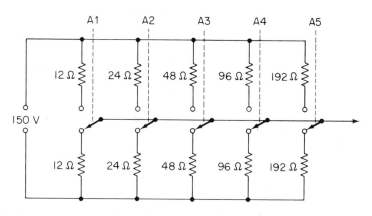

Fig. 8-28. Enlarging the Converter.

Since there is a *direct relation* between the number being converted and the resistance of the network, we can convert *directly* from the number to the analog voltage. We use the formula:

$$E_{out} = E_a \times N/M$$

Where: N is the *number* being converted and M is the *maximum* number the flip flops can contain. (These numbers are converted to decimal form).

Problem: We show five flip flops in Figure 8-28. What *analog* voltage do we obtain when we convert the binary number 11001?
Solution: The *maximum* number is 31; $2^5-1 = 31_{10}$. The *number* being converted is 25; $11001_2 = 25_{10}$.

$$E_{out} = 150 \text{ V} \times 25/31 = 3750/31 = 120 \text{ V}$$

Analog to Digital

There are many devices which change *analog* quantities of voltage, distance, volume, etc. into *digital* information. We call all of them *analog to digital* converters. When you read the time on a clock, you are making an analog to digital conversion. The hands of the clock merely show shaft *position* which is an analog quantity.

The *coding* tube provides a high speed analog to digital conversion. This tube is illustrated in Figure 8-29.

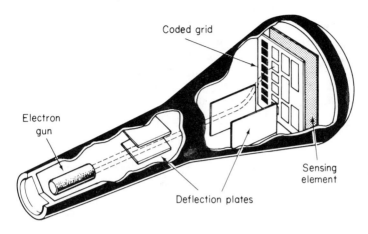

Fig. 8-29. CRT Analog to Digital Converter.

The voltage that is to be converted is connected to the *vertical* deflection plates. A linear sawtooth sweep voltage is applied to the *horizontal* deflection plates. The stream of electrons is positioned vertically on the *coded grid* so that it strikes a line of *slots* which correspond to the *amplitude* of the analog voltage. The sweep voltage causes the beam to *scan* across these slots. A *coded image* is developed on the *sensing element* which is our digital read out device.

A simple voltage *comparator* circuit may be an analog to digital converter. The general features of such a circuit are shown in Figure 8-30.

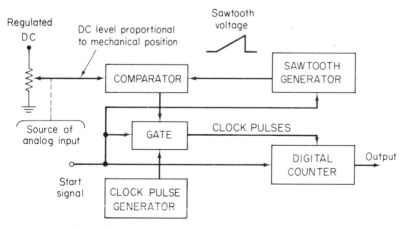

Fig. 8-30. Comparator Analog to Digital Converter.

This operation first converts a *mechanical position* to a voltage level, then converts the voltage level to a *digital* count. The turning of some mechanical device positions the *arm* on the potentiometer. Our resulting dc voltage level on the arm is *proportional* to the mechanical position. This voltage level is coupled into the *comparator*. The *start* signal enables a gate circuit, a digital counter, and a sawtooth generator. The sawtooth generator produces a *linear* sawtooth voltage which is fed into the *comparator*. The output of the comparator keeps the gate going as long as there is a *difference* between the instantaneous value of the sawtooth and the level of the dc. When these two voltages reach the *same* amplitude, the comparator output *ceases* and cuts off the gate.

As long as the gate is *on*, clock pulses are *gated* into the digital *counter*. The count when the gate shuts off is *proportional* to the level of the analog voltage.

A shaft position can be converted *directly* to a binary number by using a *binary coded disk*. A binary coded disk is shown in Figure 8-31.

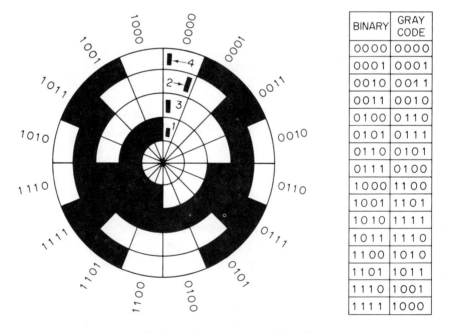

BINARY	GRAY CODE
0000	0000
0001	0001
0010	0011
0011	0010
0100	0110
0101	0111
0110	0101
0111	0100
1000	1100
1001	1101
1010	1111
1011	1110
1100	1010
1101	1011
1110	1001
1111	1000

Fig. 8-31. Coded Disk Analog to Digital Converter.

The disk is secured to a shaft whose rotational *position* we wish to convert into *digital* information. The small rectangles are *brushes* which make contact on the wheel. Each brush produces either a one or a zero. The *dark* areas represent *conducting* material, and the *white* areas are *insulated*. A brush on the conducting surface produces a *one*; on the insulating surface, it produces a *zero*.

This type of converter has some *ambiguity* because of the lack of precise changeover at the division between insulator and conductor. The code used here is called Gray code or reflected binary. The use of *Gray code* instead of straight binary is calculated to reduce ambiguity error. Why? Because the Gray code changes *only one* digit when it goes one number position in either direction. Notice also that the brushes are *offset*. This makes the readout a bit more reliable by making it impossible for *more than one* brush to give a false reading at any given time.

8-6 ADDERS

The *adder* is used in our arithmetic section of the computer not only to accomplish addition but also to *compare* numbers, *multiply*, *divide*, and even *subtract*. All these and other functions are performed through variations of the add process.

Half Adder

The half adder adds two digits and provides a sum and a carry. We had this function in our exclusive OR circuit without regard to the carry. The logic diagram is illustrated in Figure 8-32.

The half adder alone serves little useful purpose, but we use two of them to construct the *full adder* which performs our arithmetic. Input A and input

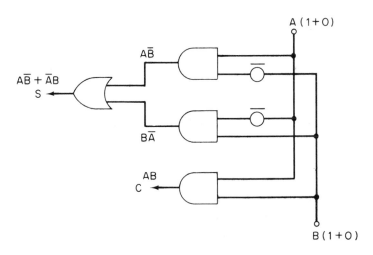

Fig. 8-32. Half Adder.

B come from corresponding bits of two registers. Let's get this very clear; the *two* numbers being added are in separate registers (A and B), and we add *one bit* from each number. This addition produces an output, S, which is the sum of the two binary bits. A third AND gate in the diagram provides a second output, C, which is the carry bit. When A and B are both *ones*, the carry bit is a *one;* otherwise, it is a zero. When A is a *one* and B is a *zero*, or vice versa, the sum is a *one*. At all other times the sum is *zero*. Figure 8-33 is the truth table for the half adder.

ADDEND	A	0	1	0	1
AUGEND	B	0	0	1	1
SUM	S	0	1	1	0
CARRY	C	0	0	0	1

Fig. 8-33. Half Adder Truth Table.

When *both* inputs are *zero*, both outputs are *zero*. When the two inputs are *different*, S = 1 and C = 0. When both inputs are *one*, S = 0 and C = 1.

Full Adder

We actually need to combine *two* half adders for the proper addition of two binary bits and the input carry bit at the same time. The truth table for a full adder is shown in Figure 8-34.

ADDEND	A	0	1	0	1	0	1	0	1
AUGEND	B	0	0	1	1	0	0	1	1
CARRY-IN	C_I	0	0	0	0	1	1	1	1
SUM	S	0	1	1	0	1	0	0	1
CARRY-OUT	C_O	0	0	0	1	0	1	1	1

Fig. 8-34. Full Adder Truth Table.

Constructing circuits to handle the bit that is carried in complicates things considerably. Figure 8-35 is a variation of the full adder.

The full adder can be constructed with *fewer* stages but *not* with fewer

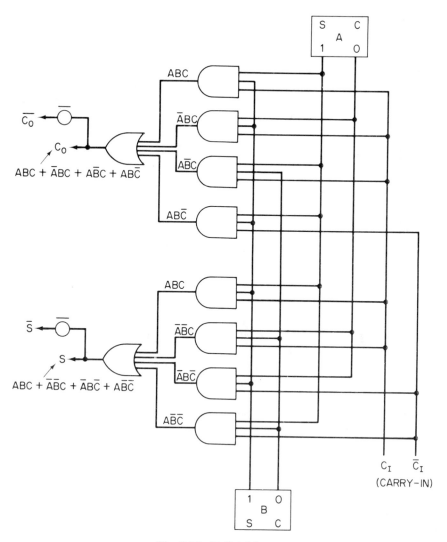

Fig. 8-35. Full Adder.

types of stages. The *carry-in* digit, either a one or a zero, *initiates* the addition. If this is the least significant bit position, a carry-in of zero is generated by the control section. The *carry-out* from this adder is the *carry-in* to initiate the addition for the next higher order column.

The sum of the two bits plus carry-in replaces the content of the A flip flop. When the entire process is completed, the sum of the two numbers has replaced the addend in the entire A register.

Serial Adder

When time is not a pressing problem, one full adder can be used to perform most of our arithmetic functions. To accomplish this, we use *shift registers* to hold the addend and augend. We *add* the LSBs and *shift* both numbers to the right; *add* the next higher order bits (now in LSB position) and *shift* again. One addition and one shift are accomplished for *each bit* in the computer word. This type of arrangement is illustrated in Figure 8-36.

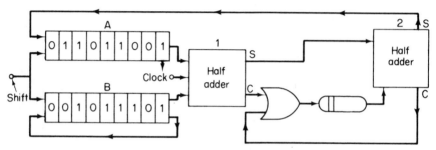

Fig. 8-36. Serial Addition with Shift Registers.

Each column addition requires a time equal to one clock *interval*. On the first clock pulse, the LSBs are *added* in HA1 to produce a sum and a carry. Using the numbers shown in the registers, we have $1+1$ for a sum of 0 and a carry of 1. The carry pulse is to be added to the next higher order column, so we *delay* it for one clock *interval*. HA2 receives S $= 0$ and C $= 0$ and produces an output of S $= 0$ and C $= 0$. The shift pulse moves *all* bits in both registers one position to the *right*. The B register is performing an *end around* shift; the LSB comes around to the MSB position. The A register is performing an *end off* shift. The LSB of the A register drops out and the MSB position is vacated to be replaced by the *sum* from HA2.

The *second* clock pulse causes the two zeros to be added in HA1. This sum (0) reaches HA2 at the same time that our *previous* carry (1) emerges from the delay line. HA2 receives S $= 0$ and C $= 1$ and produces S $= 1$ and C $= 0$. The *second* shift pulse moves the register's contents one more position to the right. The *second* bit is lost from the A register and the second sum (1) moves into the MSB position. The second bit from the B register (0) moves around to assume the MSB position.

This action *repeats* with each clock pulse until we have performed an addition on *each column* of binary digits. At that time, the content of the A register is the *sum* of the two *original* numbers. In this case, that sum is 100110110. We performed *nine* individual additions and shifted both registers *nine* times. The orginal B register contents shifted one complete *circle* to end in the starting position.

This method of performing arithmetic functions puts the burden on the *programmer* for all operations except addition. He can cause the subtraction, multiplication, and division functions to be accomplished by programming the proper *number* of ADD instructions along with some other *discrete* operations. An arithmetic element of this type may be found in *trainers* and in some *small scale* systems but *not* in a large scale system. (Scale here refers to capability, not to physical size).

Parallel Adder

A more *common* adder arrangement is shown in Figure 8-37.

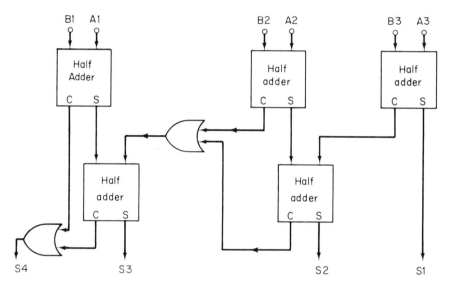

Fig. 8-37. Parallel Adder.

With a parallel adder of this type, *all* the additions take place *whether or not* the ADD is desired. All the sums are available and are held ready for use. One single clock pulse will *remove* the original contents of the A register and *replace* them with the new number. Some finite amount of time is required to complete an addition in the parallel adder, but since the addition takes place *automatically* each time that numbers are loaded into the registers, there is *no* appreciable waiting time. The reuslt is normally ready and waiting by the time the processor is ready for it.

When our computer word is in excess of 30 bits it is *essential* that we use parallel adders. Up to 30 bits per word, a serial adder is usually satisfactory, provided we have a full adder for each column of digits.

Multiplication is performed by a *series* of additions. That is, we add the contents of A and B, put the sum in A, then add A and B again. This is repeated for each bit in the computer word. When we wish the product of 5×5, we can obtain it in *two* ways; $5 \times 5 = 25$ or $5+5+5+5+5=25$. The computer uses the *second* method.

Most computers perform subtraction by *complementing* the subtrahend and *adding* the result. Division is then simply a *repetition* of the subtraction process. However, some computers *do use* subtractors.

8-7 SUBTRACTORS

There is *little* difference between adders and subtractors; both are composed of a series of gating circuits. The *adder* takes the sum of flip flop contents and carries to produce a sum and a carry. The *subtractor* takes the difference between flip flop contents, combines this with borrows, and produces a difference and a borrow.

Half Subtractor

Figure 8-38 shows the half subtractor along with its truth table.

Notice that $D = A\bar{B} + \bar{A}B$. This is the same exclusive OR function that we had for the sum in the half adder. The borrow signal, X, occurs *only* when

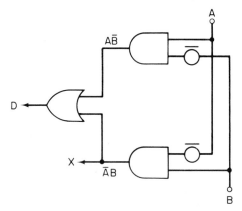

MINUEND	A	0	1	0	1
SUBTRAHEND	B	0	0	1	1
DIFFERENCE	D	0	1	1	0
BORROW	X	0	0	1	0

Fig. 8-38. Half Subtractor.

the minuend is *zero* and the subtrahend is a *one*; $X = \bar{A}B$. When a borrow occurs, a one is *subtracted* from the minuend of the next higher order column.

Full Subtractor

The full subtractor is a combination of *two* half subtractors. The truth table is shown in Figure 8-39.

MINUEND	A	0	1	0	1	0	1	0	1
SUBTRAHEND	B	0	0	1	1	0	0	1	1
BORROW-IN	X_I	0	0	0	0	1	1	1	1
DIFFERENCE	D	0	1	1	0	1	0	0	1
BORROW-OUT	X_O	0	0	1	0	1	0	1	1

Fig. 8-39. Full Subtractor Truth Table.

From this truth table we can construct our *equations* for difference and borrow.

$$D = A\bar{B}\bar{X}_I + \bar{A}B\bar{X}_I + \bar{A}\bar{B}X_I + ABX_I$$

$$X_0 = \bar{A}B\bar{X}_I + \bar{A}\bar{B}X_I + \bar{A}BX_I + ABX_I$$

A *parallel* subtractor for taking the difference of *two* (three bit) numbers is illustrated in Figure 8-40.

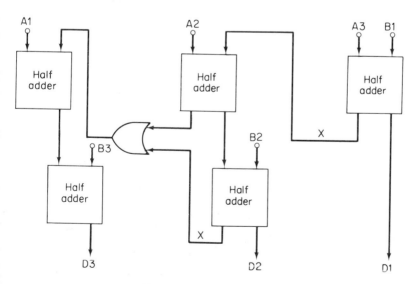

Fig. 8-40. Parallel Subtractor.

A serial subtractor can be arranged identically to the serial adder in Figure 8-36, or we can have a full subtractor for each column of digits.

Review Exercises

1. (a) What is a functional component?
   (b) Give three examples.

2. What is a register?

3. Distinguish between a storage register and a shift register.

4. In order to accomplish a proper ones side transfer between registers, what must be done to the receiving resgister prior to the transfer?

5. The bit positions in a certain register are numbered A1, A2, A3, and A4. Identify the LSB and MSB.

6. Redraw logic diagram of Figure 8-2 to accomplish the forced feed transfer with a minimum of logic.

7. What is the advantage of a forced feed transfer?

8. (a) When data is being received from a telephone line, what type of register is likely used to accumulate this data?
   (b) Why?

9. Why are parallel transfer registers normally found in the central computer?

10. What provision is made to avoid loss of data when performing multiplication or division?

11. (a) What operation must be performed on the result of a multiplication or division prior to storage?
   (b) Describe how this is done.

12. Draw a block diagram to illustrate a double, right, arithmetic, and end around shift.

13. Identify the type of shift illustrated in Figure 8-41.

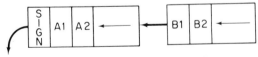

Fig. 8-41. What Type of Shift?

14. How does the computer differentiate between a positive number and a negative number?

15. How does the computer determine exactly what type of shift to perform?

16. Redesign the counter in Figure 8-11 to a serial down counter.

17. A four stage down counter starts with all flip flops set. How many input pulses are required to reduce the count to 0000?

18. List two distinguishing features of the parallel counter.

19. (a) What is the total number of possible conditions for an eight stage counter?

(b) What is the maximum count it can contain?

20. A seven stage up counter has a count of 1011010_2. After 200_{10} pulses are applied, what is the new count and how many cycles were completed?

21. A ten stage up counter is counting clock pulses with intervals of 0.5 μs. How much time is required for each counter cycle?

22. Draw an AND gate detector that will detect the number eight in a four stage register.

23. What numbers are detected by the gates in Figure 8-42?

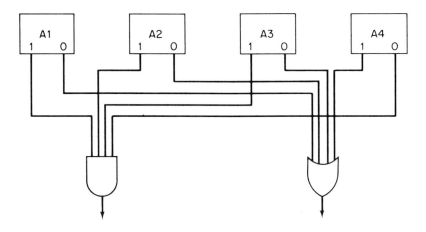

Fig. 8-42. What Numbers are Detected?

24. Write the Boolean equation for both outputs of Figure 8-42.

25. Draw a schematic of a diode matrix on the output of two flip flops. Connect an AND gate to detect a binary count of 10 and an OR gate to detect $\overline{01}$.

26. How many diodes are needed in a common matrix which detects all possible conditions of an eight stage counter?

27. How does a treed matrix differ from another type matrix?

28. Using the converter in Figure 8-28, a binary configuration of 10101 will produce what value of analog voltage?

29. Name three methods of converting analog quantities to digital counts.

30. What conditions will cause a half adder to produce a sum of one? A carry of one?

31. (a) Draw the logic for this equation: $ABC + \bar{A}\bar{B}C + \bar{A}B\bar{C} + A\bar{B}\bar{C}$. (b) Identify the logic you have drawn.

32. Explain how a computer can multiply, subtract, and divide by using adder circuits.

33. What are the principal differences between an adder and a subtractor?

chapter nine

computer memory

Any type of device which can be used to store and retrieve information is, in a sense, a *memory* device. We have a wide range of memory devices, and all of them have been used in computer systems. In this chapter we will examine several types of storage devices and the methods of storing and retrieving information. Finally, we will take a close look at the organization, timing, and control of a ferrite core *central* memory.

9-1 TYPES OF STORAGE

There are *four* general methods of storage: mechanical, electronic, optical, and magnetic. Any of these types can be organized to form a computer memory.

Computer memories are divided into *two* categories: file memories and dynamic memories. The dividing line between these two categories is steadily growing thinner. At least one company has a random access, central memory which uses magnetic cards as the storage device.

Mechnical Storage Devices

There are two types of *mechanical* storage devices in popular use. These are *punched cards* and perforated *paper tapes*. Samples of both are shown in Figure 9-1.

The punched card is one of the most successful methods of communication between man and machines. It was the *first* storage device and still enjoys great popularity. Today, it is big business just manufacturing the standard card.

The standard 80 column card was developed by IBM and is the card most widely used in modern equipment. The dimensions are 7 3/8″ × 3 1/4″ × 0.0065″. It has 12 rows for punching information. The standard *Hollerith* code is the most common recording code; however, there are other codes in use. The Hollerith code is punched into the card in Figure 9-1b. The complete code is repeated in Figure 9-2.

Sooner or later, practically all program information is recorded on punched cards. The information is recorded by a card punch. The punch may be a *manually* operated machine which is controlled by a typewriter keyboard, or it may be a machine which is operated by *computer* control. Information is retrieved from the card by a card reader. We will have more to say about these machines in the chapter on input and output.

The punched card is a durable, reliable method of storing information in a file memory. It is also a handly method of exchanging information with our central computer.

Perforated paper tape is also a convenient method of mechanical storage. Figure 9-1C shows a reel of this tape, and 9-1D is a section of the tape with data recorded. The large dots are *holes* which have been punched for data. This particular tape has information recorded in a five channel code. The channels are parallel along the length of the tape. One *column* across the tape is used for recording each character. This code uses four channels to record the common characters, and one channel for special characters.

The row of small, evenly spaced holes which runs lengthwise on the tape is for the *star-wheel* drive mechanism. Points on the star-wheel fit through these small openings and *move* the tape through the punch when recording and through the reader when reading.

Paper tape is a continuous recording medium which provides unlimited storage for our file type memories. Most keyboard input/output devices contain a combination paper tape reader and punch. In this case, *typed* information can be automatically *punched* into the tape, and when we run a

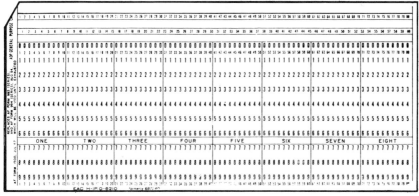

(a) Standard Card

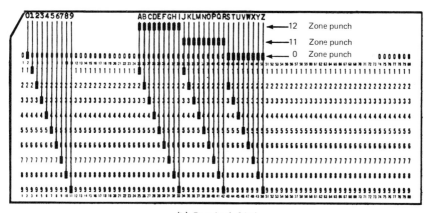

(b) Punched Card

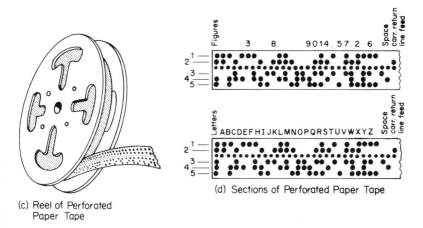

(c) Reel of Perforated
Paper Tape

(d) Sections of Perforated Paper Tape

Fig. 9-1. Samples of Mechanical Storage Devices.

Character	Punched— Card Code		Character	Punched— Card Code
0	0		—	11
1	1		J	11 − 1
2	2		K	11 − 2
3	3		L	11 − 3
4	4		M	11 − 4
5	5		N	11 − 5
6	6		O	11 − 6
7	7		P	11 − 7
8	8		Q	11 − 8
9	9		R	11 − 9
@	8 − 2		⌐ (Tab)	11 − 8 − 2
=	8 − 3		S	11 − 8 − 3
;	8 − 4		*	11 − 8 − 4
≡	8 − 5		<	11 − 8 − 5
&	8 − 6		#	11 − 8 − 6
'	8 − 7		⊔	11 − 8 − 7
+	12		⟋	0 − 1
A	12 − 1		S	0 − 2
B	12 − 2		T	0 − 3
C	12 − 3		U	0 − 4
D	12 − 4		V	0 − 5
E	12 − 5		W	0 − 6
F	12 − 6		X	0 − 7
G	12 − 7		Y	0 − 8
H	12 − 8		Z	0 − 9
I	12 − 9		⌐	0 − 8 − 2
∩	12 − 8 − 2		,	0 − 8 − 3
.	12 − 8 − 3		(	0 − 8 − 4
)	12 − 8 − 4		>	0 − 8 − 5
%	12 − 8 − 5		:	0 − 8 − 6
?	12 − 8 − 6		e	0 − 8 − 7
"	12 − 8 − 7			

Fig. 9-2. Hollerith Code.

punched tape through the machine, it will automatically *operate* the type-writer.

Here we have illustrated a five channel code, but the code used in any case is determined by the designer of the (keyboard) reader/recorder. He builds the necessary circuits to interpret the code which he has elected to use. All codes are similar to the one described.

Electronic Storage Devices

Today we have flip flops, flip flop latches, and storage tubes which are used to *electronically* store information. Solid state flip flops and latches have recently become very *reliable*, relatively low cost methods of storage. We find them in many *auxiliary* memories, and in a few systems, they are used as storage devices in the central memory. Metal oxide semiconductors used in integrated circuits provide nonvolatile storage. This means that power can be removed and reapplied without the flip flop (or latch) changing states, and marks an important forward step in memory technology.

The coded tube which we discussed in chapter 8 is another method of storing information electronically. There are many variations of the storage tube which provide limited memories of a temporary nature.

The ovonic switching device is a late development in experimental storage devices. This is an *amorphous* semi-conductor composed of a special type of *glass*. A bead of ovonic material has *two* stable states that can be reversed by electric signals. It is nonvolatile in nature and has a very rapid cycle for reading or writing. It presently requires a separate semiconductor device for each bit stored. This poses an economic limitation, but the problem will probably be solved in the near future.

Optical Storage Devices

Optical storage is a fairly recent development in computer memories and has severe limitations at this time, but should become practical in a few more years. The storage medium is a *photographic* plate on which we record information as either a real or a holographic image. The information is retrieved by painting the accessed areas with a *laser* beam.

The two major objections to the optical memory system are *cost* of laser apparatus and *inflexibility* of the stored information. The cost can be expected to decrease, and it is reasonable to assume that technological advances will produce the necessary flexibility. A permanent memory, such as we have with our present optical systems, may have definite applications, but they are extremely limited in our present computer systems.

Magnetic Storage Devices

Tapes, drums, disks, thin film spots, magnetic domains, magnetic cards, and ferrite cores are some of our storage devices which utilize *magnetic* storage. With the exception of magnetic domains, all of these are practical, proven, and enjoy wide application. Among these, we find drums, disks, thin film spots, cards, wires, and cores as the storage devices in *central* memories. Tapes, disks, and drums are very popular for *file* type memories. Tapes and disks provide *unlimited* capacity for our memory library. Disks and drums are also very flexible as *dynamic* storage devices. They are frequently used in conjunction with display systems.

Magnetic domain storage is an *experimental* method of storing information. We find these magnetic domains existing in thin sheets of garnet or orthoferrite; we sometimes call them magnetic *bubbles*. The domains can be selectively formed, detected, and erased by magnetic fields.

Information is accessed in this type of memory in serial form. This creates

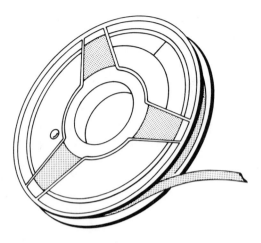

(a) Reel of tape

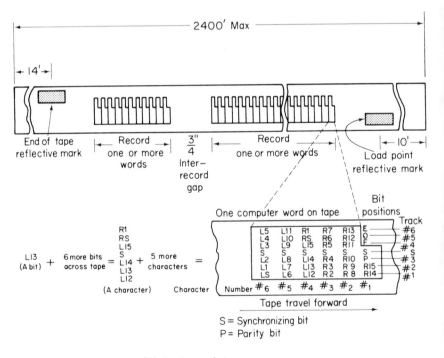

(b) Sections of tape

Fig. 9-3. Magnetic Tape.

a relatively *long* access time for particular bits, but even with this disadvantage, it promises to become a great asset to the area of *file* type memories.

Tapes, drums, and disks use essentially the same techniques in storing, retrieving, and erasing information. We form all three of these devices by laying a thin ferromagnetic surface over an insulating substrate. The names (tape, drum, and disk) are descriptive of the shape of the substrates. Information is recorded by *magnetizing* small areas of the magnetic surface. *Write* heads record information, and *read* heads retrieve this information.

Figure 9-3 shows a tape reel and an enlarged section of magnetic tape.

Magnetic tape is one of our principal input/output devices in modern equipment. This is an easy, convenient method of recording *large* quantities of information for entry into the central computer. A reel provides up to 2400 feet of storage area. Tapes may be used in a *dynamic*, on line operation to furnish data as the computer needs it. They can also be *accessed* by the computer for storage of large blocks of output data.

The drawing in Figure 9-3B represents a *seven* track (channel) recording of an *alphanumeric* code. In this case, six channels are used for the character, and the seventh (s) is for sychronization. A section of this tape is enlarged in Figure 9-4 to illustrate the *polarities* of the magnetic fields.

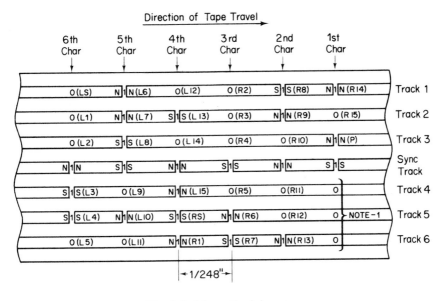

Fig. 9-4. Magnetized Areas.

Notice that some spots are south and others are north. This is a method of recording which we call "*non return to zero*" (*NRZ*). This method of recording *reverses* the flux to record a *one* and leaves the flux unchanged to record a zero. This is illustrated by a flux wave shape in Figure 9-5.

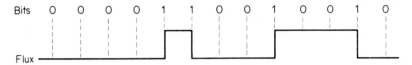

Fig. 9-5. NRZ Method of Recording.

Each *column* contains seven magnetized areas; we have shown only the *changes* in flux polarity. The areas with zeros are still magnetized, but flux is in a *constant* direction. For instance, track 1 in Figure 9-4 has a south polarity until we reach the first character point. Each column in this area has a recorded bit. At the character point, we *change* from south to north and record a *one*. In the next column, we record a zero by *not changing* the polarity. The same bit and flux pattern is shown in Figure 9-5, but the wave shape represents only *one* track on the tape. It *is not* necessary to erase; new information can be recorded over the old.

Tapes are run at very *high* speeds, and any tape record may be addressed. This makes retrieval time reasonably short, but access is still *serial*. This renders it unsatisfactory for central memories.

Magnetic disks have been used to reduce the serial access problem which is inherent in tapes. A disk drive and a magnetic disk pack are shown in Figure 9-6.

We encounter some disk drives which house a *specified* number of disks. These disks resemble large phonograph records with recording surfaces on both sides. A disk drive module with *fixed* disks generally has about 25 of these double surface disks. Each disk is approximately *two feet* in diameter with 500 or more tracks (channels) on each side.

There are *other types* of disk drives which provide greater *flexibility*. They use *removable* disk packs as shown in Figure 9-6. These disk packs consist of *six* or more disks which are fixed on a vertical shaft. The outer surface of the top and bottom disk are not used. The other 10 surfaces are used for recording information. Read-write *heads* resemble teeth on a comb. There is a head assembly for *each* surface.

Information on a disk is *addressable* to a track. This means that we have random access to any track, and access time to any point of a track can be no more than the time of one disk revolution. The disk can be used for *central* memories, and the removable pack provides us with unlimited *file* storage. The process of recording and reading information is similiar to that for magnetic tapes.

Magnetic drums provide us with another flexible method of storage. A

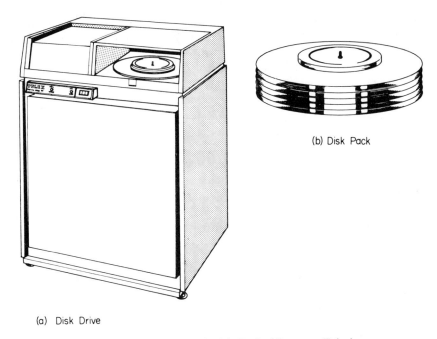

(b) Disk Pack

(a) Disk Drive

Fig. 9-6. Disk Drive and Disk Pack. (Courtesy Telex)

drum is a metal *cylinder* coated with ferromagnetic material. Drums range from two inches to four feet in *diameter* and from a quarter inch to five feet in *length*. The surface is divided into circular *bands* which we refer to as either tracks or channels. There are several bands for each inch of the drum length. A magnetic drum is represented in Figure 9-7.

Bits of information are recorded along a band in a manner similar to that described for tapes. A write head records information by magnetizing small

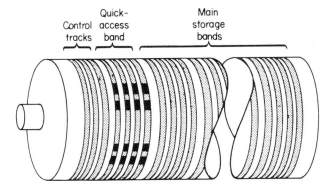

Quick-
Control access
tracks band

Main
storage
bands

Fig. 9-7. Magnetic Drum.

areas. A read head retrieves the information by detecting the flux. Tracks are addressable for direct access, but information within a track is in serial form.

The drum *rotates* at a high speed which causes each area on a track to be accessible in a short time. Several sets of read write heads may be placed at various points around a track to further *reduce* access time.

Drums are used in many ways, and *most* systems contain from one to several. They provide *dynamic* storage for data which are used frequently, *temporary* storage for data being transferred through, or *semipermanent* storage for certain types of programs. When system use can tolerate relatively slow access time, the drum is a very satisfactory *central* memory.

Thin film spots enable us to construct a fast, random access memory that is widely used for *auxiliary* storage. Thin film has not become as popular as central memories because of the relatively high *cost* per bit of storage area.

The film is constructed by depositing small spots of nickel-iron alloy on a glass substrate. The spot is *magnetized* during deposition, and current passing near the spot will *shift* the magnetic axis. An axis in one direction constitutes a one, and a different direction constitutes a zero. We can store one binary bit on each thin film spot. Retrieval is accomplished by passing read current in the opposite direction. A section of thin film is illustrated in Figure 9-8.

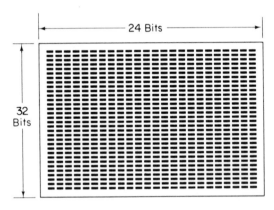

Fig. 9-8. Thin Film Section.

This section of thin film can be used to store 32 *words* of 24 *bits* each. Several sections may be combined to *increase* word length or word capacity, or both.

Ferrite cores have consistently proved to be the most *desirable* type of storage for *central* memories. The storage device is a small ring composed of ferromagnetic material. The ingredients are iron oxide, manganese, oxides, and an organic binder. These materials are mixed, bound, and baked in a kiln to form a brittle, ceramic-type doughnut. We call this tiny doughnut a ferrite core.

The ferrite core has a *low* reluctance and a *high* retentivity which makes for easy magnetization and indefinite retention of information. A *straight wire* threaded through a core is sufficient for storing or retrieving information. The core stores a single binary bit when it is magnetized by current passing through the opening. Current in one direction creates flux in the core of a particular polarity. *Reversing* the current removes the flux and establishes a flux of the *opposite* polarity. We select flux in one direction to represent a one and in the opposite direction for a zero. After the core is magnetized the first time, it can *never* be neutral again. It *always* contains either a one or a zero. Figure 9-9 illustrates the function of a ferrite core.

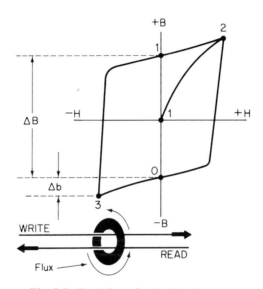

Fig. 9-9. Function of a Ferrite Core.

When we apply a specified level of current on the *write* line, we are applying a magnetizing force in the direction of $+H$. This creates *flux* inside the core in the direction indicated by the arrows. On the hyestersis loop, flux *rises* along the line from 1 to 2. When write current ceases, the magnetic flux decreases to the upper point on the $+B$ line. We have stored a *one*. The distance from this point to the H line represents the *residual* magnetism which remains after current is terminated in the line. This is the flux which holds the core in the *one state* for an indefinite period of time.

We send current through the *read* line when we wish to retrieve the one bit. This is a magnetizing force toward $-H$. When our current reaches the *switching* level, the flux *reverses* direction and swings to point 3. When we terminate the read current, flux decays to the lower point on the $-$ B line. This is the flux retained for our *steady* state zero. The zero will remain until we deliberately apply *write* current and switch the core back to the one state.

ΔB shows the *difference* in flux between a steady state one and a steady state zero. Δb shows the difference between the *peak* flux when switching takes place and the *steady* state flux of residual magnetism.

This relatively *square* hysteresis loop marks one of the highly desirable characteristics of the core. A loop of this type indicates that the core *is not* likely to switch accidentally, can retain information *indefinitely*, and is *non-volatile* (retains information with power off). When switching is induced, the *rapid* change of flux produces a strong output signal.

One line through a core for read and one for write, are *not* sufficient when we group them with other cores. We actually need two lines as indicated in Figure 9-10 for either read or write, but the *same* two lines can be used for *both* read and write by properly routing the current through the lines.

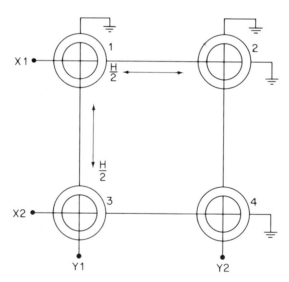

Fig. 9-10. Core Selection.

This is called a coincident current memory. When an address is decoded, it *selects* one core on a memory plane. The decoder *enables* one X line and one Y line and causes a current in each of those lines. This current on each line is *half* enough to switch a core. The two lines are common to one core on each plane. Figure 9-10 represents part of a plane. If we have placed *half* switching current on X2 and half on Y2, core 1 has no current, cores 2 and 3 have *only* half current and *do not* switch; core 4 has *full* current, and it *switches* to the opposite state. If currents toward ground are for writing, currents in the opposite direction, on the same lines, are for reading. (H/2 = half switch current).

A core memory generally has a *plane* of cores for *each* bit in the computer word. This permits storage of a word in a *vertical* position with one bit per plane. This is illustrated in Figure 9-11.

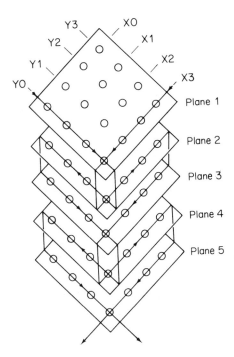

Fig. 9-11. Word Orientation.

In this drawing, we have inserted the X and Y lines for memory address X3-Y0. Notice that these lines *intersect* through *one* core on each plane. Currents in the direction of the *arrows* select and write into the cores of the word. Currents against the arrows *reverse* the flux to read out the information.

We need two *more* lines for each core in order to have a *practical* storage device. Figure 9-12 illustrates a core threaded with all the essential lines.

In addition to the X and Y lines which we had previously, this drawing shows a *sense* line and an *inhibit* line. The sense line threads the core *parallel* to the X line. The inhibit line is *parallel* to the Y line.

The direction of current on the X and Y lines is controlled by *diodes* which are numbered from 1 through 8. Diodes 2 and 5 are enabled by Y *drivers* (YD) while 1 and 6 are enabled by Y *switches* (YS). Diodes 4 and 8 are enabled by X drivers (XD), and 3 and 7 are enabled by X switches (XS).

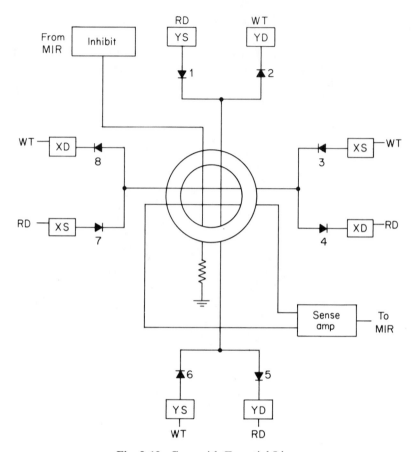

Fig. 9-12. Core with Essential Lines.

Diodes 2 and 6 work together to supply a path for *half* switch current from YD through YS. This path is completed when we wish to *write* (WT) information. Also during write, we *enable* diodes 8 and 3 to provide a path for *half* switch current from XD through XS. If we desire to write a *one, only* these two lines (X & Y) carry current. On the *write* command the core is *always* in the zero state. The *coincidence* of the two half currents in the core will *reverse* the flux and record a one.

When we desire to record (WT) a *zero*, the problem is to *avoid* writing a one; the zero is already there. We still have *half* switch current on both X and Y lines, so we need another current to *cancel* the effect of one of these. The *inhibit* line is enabled by *half* switch current when the bit being written is a *zero*. The inhibit current is from ground to the inhibit block. Y current and inhibit current are *equal* in amplitude, and pass through the core in *opposite* directions. This cancels the effect of the current in the Y line. X current *is not* affected, but X current alone is *insufficient* to switch the core. The flux remains in the zero direction; the zero, which was already there, has been *preserved*.

During the *reading* operation (RD), we *reverse* the flux of all cores which contain *ones*, but do nothing to the cores which contain *zeros*. This still requires *full* switch current, half on the X line and half on the Y line. Diodes 4 and 7 complete the X path from XD through XS. The Y path is from YD through D5, D1, and YS. Notice that both of these currents are in *opposite* directions from the write currents. If the core contains a *one*, the flux *reverses* to the *zero* state. The *changing* flux *induces* a pulse into the *sense* line. The sense amplifier amplifies and shapes this relatively weak pulse into a *standard* pulse to represent a one. If the core contains a *zero*, the X and Y currents *aid* the magnetic flux which is already there; *no* switching occurs, and there is no induced pulse on the sense line. The pulse amplifier interprets this lack of a signal as a *zero* and leaves the associated flip flop in the memory information register in the clear state.

The memory information register (MIR) receives the bits from the sense amplifiers when a word is read from core storage. There is a separate core, sense amplifier, and MIR flip flop for *each* bit of a complete word. This register also holds the new word which is being written into core storage. In this case, each of the *clear* flip flops in MIR enables the *inhibit* circuit for the associated core to retain the core in a *zero* state. The set flip flops in MIR *do not* enable their inhibit circuits, and the associated cores store one bits.

There is *one* inhibit driver for *each* plane in a core memory, and the inhibit line from this driver is threaded through *each* core on that plane. The *half* drive current is present in this line only when writing a *zero*. One method of threading these lines is illustrated in Figure 9-13.

In this drawing, we have deliberately changed terminology from line to *winding*. Many texts refer to these straight lines as if they were windings on a solenoid. Taking note of this fact here may help to avoid confusion later.

We also have *one* sense amplifier for *each* plane, and we thread the sense line from that amplifier through *each* core on that plane. One method of threading the sense line is illustrated in Figure 9-14.

We have been dealing with very small, representative core matrices and will continue to do so, but we need to remind ourselves from time to time that a core matrix can be *very large*. A matrix for a core memory plane can be purchased with wiring already installed. The standard size is 64^2, i.e., 64 cores wide and 64 cores deep. Such a matrix is illustrated in Figure 9-15.

The individual cores are small; outside diameter is 0.083 inches and inside diameter is 0.05 inches. They are also mounted very close together, but $64^2 = 4096$ individual cores. In addition to this, we need a separate plane for each bit in the computer word. This means that a memory for a large scale system can have up to 80 planes in a stack.

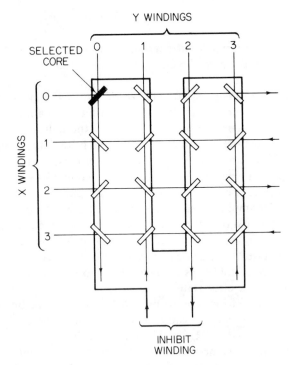

Fig. 9-13. Pattern of One Inhibit Line.

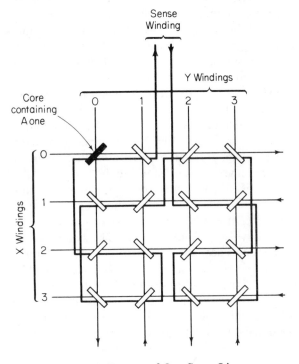

Fig. 9-14. Pattern of One Sense Line.

Fig. 9-15. Matrix for Core Memory Plane.

9-2 CENTRAL MEMORY

We have examined a variety of storage devices. Now we will select a suitable device and organize a *central* computer memory. Since the ferrite cores are still the most popular storage media for that purpose, we shall use cores in our *sample* memory.

The central memory is the most *active* section of a computer system. *All* information that passes to and from the central computer (both program and data) passes through the central memory. The memory must provide *storage* for this information and hold it available at all times.

Functional Analysis

The *vital* areas, with the exception of timing, are shown in the block diagram of Figure 9-16.

We have constructed our sample memory for a *capacity* of 4096_{10} words. Each word has 15 significant bits plus one bit for the sign and one bit for parity. (The parity bit is an extra bit used as a quick check on the validity of a store-retrieve operation.) Our core *array* is a stack of 17_{10} planes, and each plane is 64 square (4096_{10} cores/plane). The word layout is *vertical* from top to bottom as illustrated in Figure 9-17.

It is customary to express addresses in *octal* numbers. Notice in this drawing we labeled the X and Y lines in octal. 0-77 octal is the same as 64 decimal. We have X lines 0-77 and Y lines 0-77. This provides 4096_{10} *intersections* on each plane. Each bit position of our computer word has a given plane with the sign bits on the top plane and the parity bits on the bottom plane.

We have 64_{10} variables in a group of 6 binary bits. This means that we need 12 bits of *address:* six for X and six for Y. This gives us a *range* of addresses from 0000 to 7777 (octal).

In Figure 9-16, the 12 bit memory address register has been *cleared* to 0000. Memory serves all input/output devices as well as the arithmetic section (computer) of the central processor. If we have *several* active input/output devices, we will need a *priority* system to establish the order of memory access. When access is granted, the 12 bit address is furnished by the *requesting* device. The *load* signal gates the address into the MAR. The address bits will be held here throughout the memory cycle.

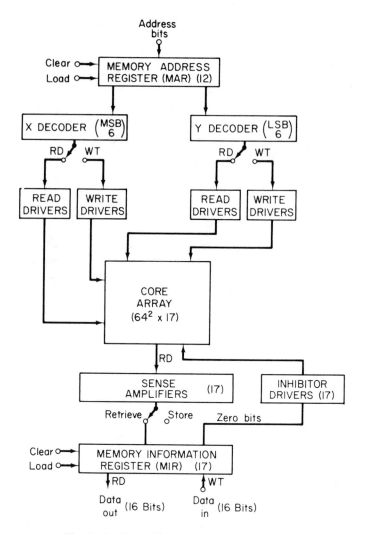

Fig. 9-16. Central Memory Functional Diagram.

A memory *cycle* is a series of events which start when the address is loaded into MAR. It includes removing a selected word from the core array and writing the same word, or a new word back into the same location.

The X decoder examines the six *most* significant bits and selects an X *driver*. The RD-WT switches are actually gates which are activated by the *timing* in our memory cycle. We have 64_{10} read drivers and 64_{10} write drivers. One of these will be selected. Their numbers are 00 to 77 (octal) for both RD and WT. The number selected will be the one that *matches* the octal configu-

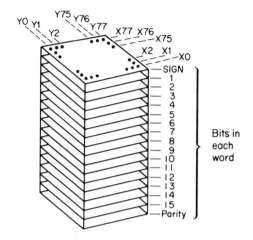

Fig. 9-17. Core Array.

ration of the six MSBs of the address. Assuming that the address is 10101001111_1 (5237_8), the X decoder will select read driver 52.

The Y driver is selected by the decoding of the *six LSBs* of the address. In this case, we select Y driver 37. The 17 cores in word location 5237 will *each* receive *half* current from each of the selected drivers.

The cores in this word which contain *ones* will have their flux *reversed* to the zero state. The *sense* amplifiers (one for each bit) which monitor these bits will receive a pulse of current which is induced by the *change* in flux. These pulses will be amplified and shaped into standard one signals and placed into the MIR. The cores in the selected location which contained *zero* had *no* change in flux. The corresponding sense amplifiers receive *no* pulse and interpret the absense of a pulse as a *zero*. MIR was cleared earlier and the *load* signal moves all 17 bits into the register at the same time.

The information is now in the MIR and will be gated out through the data lines to the *requesting* device. After this transfer, we still have a copy of the word in the MIR. The location in the core array has been *cleared* to all zeros by the reading action. Before we terminate the memory cycle, we must *rewrite* the word from MIR back into the same location.

At a certain point in our memory cycle, the RD gates will be *deactivated* and the WT gates will be *activated*. The X and Y write drivers for the same location will be activated. The cores which correspond to *one* bits in MIR will receive *full* current. The cores that correspond to *zero* bits in MIR will receive *inhibit* current from the inhibit drivers. The *one* bits will cause a *reversal* of flux in the core from zero to the one state. The *zero* bits *inhibit* the drive current from altering the flux in the cores. These cores remain in the zero state. The end result is a *rewriting* of the same word that we removed from core storage.

Our *storage* cycle is much the same as the retrieval cycle just described. The cycle starts the same way. The location is selected and read out. The sense amplifiers pick up the bits, but the *lines* between the sense amplifiers and MIR have been *opened*. This leaves both the core location and MIR *cleared* to all zeros. The memory cycle is now about half over.

The *requesting* device now loads MIR with a 16 bit word. The read drivers are switched out and the write drivers are switched in. The *new word* is recorded from MIR to the cores in the selected location.

Notice that we have 17 bits in the MIR, sense amplifiers, and core array, but there are only 16 bits transferred between MIR and the requesting module. The difference is the *parity* bit. As a word enters MIR for storage, the configuration is checked for one bits. If the word has an *odd* number of one bits, the parity position (17) is filled by a *zero*. If the word contains an *even* number of one bits, the parity position is filled by a *one*.

Thus, all words being stored contain an *odd* parity. As these words are removed, parity is checked again in the MIR. If the word has *odd* parity it is assumed to be *correct*. If it has *even* parity, a data *error* will be signaled.

Once parity has been verified, the parity bit is *dropped*. It is intended as a memory self check. It has no other function.

Of course, a memory may be designed for *either* odd or even parity. We selected odd because it is most frequently encountered.

Timing and Control

Each step of the memory cycle is controlled either directly or indirectly by a 12 flip flop ring counter. All timing and control functions belong to the *control* section, but these functions are scattered throughout the systems. Covering the control section in a separate chapter would result in a loss of continuity. To avoid this, we will cover portions of the control section as we encounter its influence.

We have *two* types of memory cycles: retrieval cycle and storage cycle. *Each* of these cycles has *two* types of action: read and write. The retrieval cycle is frequently called a read cycle, and the storage cycle is often called a write cycle. Regardless of the names used, we have one cycle for *removing* information from storage and another cycle for *placing* new information into storage. On the *read* portion of the *retrieval* cycle, we move information from core to MIR. During the *write* portion of the *retrieval* cycle, we move information from MIR back into core and gate a copy of this information to the requesting device. On the *read* portion of a *storage* cycle we clear the cores in the selected location and load MIR with a new word from the requesting device. During the *write* portion of the *storage* cycle, we write the new word from MIR into core. Figure 9-18 is a detailed chart of the timing and control signals for *both* types of cycles.

Assuming that our clock pulses have a pulse recurrence *frequency* of 2 MHz, we will have 0.5 μs between pulses. RC on the chart is *ring counter*.

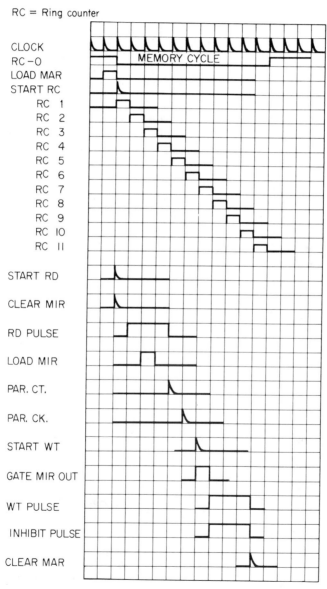

Fig. 9-18. Memory Timing and Control Signals.

The zero flip flop in the RC was set when our last memory cycle terminated. It will remain set until we gate it into action.

When a requesting module gains *access* to memory, *load MAR* moves a 12 bit address into MAR. On the next clock pulse, *start RC* gates the ring counter into action by clearing RC0 and setting RC1. Subsequent clock pulses clear the set flip flop and set the next higher order flip flop. The memory cycle *terminates* when RC11 clears and RC0 sets again.

Another pulse starts the read portion of the cycle at the same time that RC1 is set. A third pulse, at this same time, clears MIR. The negative transition of RC1, along with a clock pulse, sets RC2 and generates the *RD pulse*. The RD pulse *enables* the read drivers that have been selected by the address decoders. The information moves from the cores to the sense amplifiers, and the *load MIR* signal occurs at the start of RC3.

At this point, we have our first difference in the two types of cycles. On the *retrieve* cycle, load MIR moves information from the *sense* amplifiers into MIR. On the *storage* cycle, the sense amplifier inputs are disabled, and load MIR moves a *new word* from the requesting device into MIR.

At the start of RC5, we perform a *parity* count on the word in MIR. If we are in a *retrieval* cycle, an *even* count constitutes an error which results in a parity *alarm*. If we are in a *storage* cycle, an *even* count sets the parity bit to a one (an odd count leaves it as a zero). After assigning the parity bit, parity is checked again. If the count is *odd*, timing procedes in a normal manner; if the count is *even*, we have problems in our parity assignment circuits. An even count causes a parity *alarm*.

At the start of RC7, parity having checked properly, the *start WT* pulse starts the write portion of the cycle. At this same time, we generate a signal called *gate MIR out*. This is the signal which moves the word from MIR to the requesting device on retrieve cycles. On storage cycles this signal accomplishes nothing.

At the start of RC8, the *WT pulse* activates the selected write drivers, and the *inhibit* pulse activates the proper inhibit drivers. This combined action *writes* the word from MIR into the selected core location.

At the start of RC11, we clear MAR to prepare for the next cycle. When RC11 terminates, it sets RC0 and concludes the memory cycle.

Parity Circuits

Our parity circuit counts parity by *checking* the one bits in each flip flop. It doesn't actually count, it just *keeps track* of odd bits or even bits. One type of parity checking circuit is illustrated in Figure 9-19.

This is a parallel checking circuit. *Each pair* of flip flops feeds an exclusive OR circuit. These combinations are worked down two at a time until there are *only two* outputs for the entire register. These two outputs then feed ex-

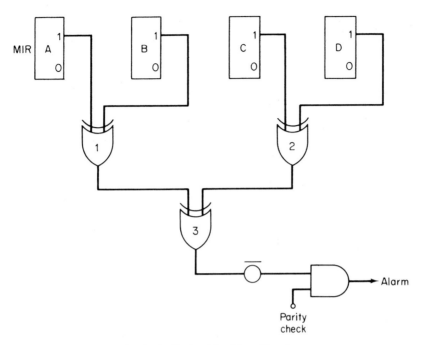

Fig. 9-19. Parity Checking Circuit.

clusive OR 3. When the register contains an *even* number of one bits, OR 3 will have a *low* output. After passing the inverter, the low becomes a *high* and conditions one leg of the AND gate. The *parity check* pulse conditions the other leg and activates the AND gate. A *high* from the AND gate triggers the parity *alarm* circuit.

Most parity circuits not only check parity, they *assign* the parity bit to give proper parity to the word being stored. This is illustrated in Figure 9-20.

This circuit clarifies the *three* steps in determining valid parity during storage.

1. Count parity
2. Assign parity
3. Check parity

The exclusive OR circuits count the parity at a specified time (not shown). The result of this count will be a *high* from OR 4 when parity is *odd*, and a *low* when parity is *even*.

Starting with the parity flip flop *clear* and all others *set*, we have a *high* from OR 1, a *low* from OR 2, a *high* from OR 3, and a *low* from OR 4. The low from OR 4 is *inverted* to a high and applied to the two AND gates. The next signal generated is *assign parity*. AND 2 is activated and triggers the parity

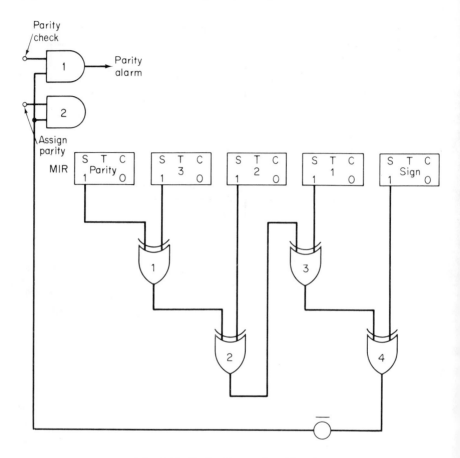

Fig. 9-20. Parity Generation Circuit.

flip flop from zero to one. Another count takes place with these results: a *low* from OR 1, a *high* from OR 2, a *low* from OR 3, and a *high* from OR 4. The high from OR 4 is *inverted* to decondition both AND gates.

This was a proper parity check. The check parity pulse finds AND 1 *deconditioned* and there is no output to cause an alarm. However, if our *second* count had still revealed *even* parity, AND 1 would be activated by the *parity check* signal. This would have set off the parity alarm.

Increasing Memory Capacity

The *capacity* of our central memory is not limited to 4096 locations. Other modules of like capacity may be added on. One system uses 16 modules of 4096 words each. How do we make this compatible with our 12 bits of address? First, we *are not* limited to 12 bits of address. Some machines use more address bits, some use less. Second, we can use *indirect* addressing, and third, we can *modify* our address by adding a predetermined number. In indirect addressing, the computer instruction gives an address of a core location whose content is the actual data address.

Suppose that we have two memory modules of 4096 words each. The addresses of the first module are 0000 through 7777. The addresses of the second module start with 7777 + 1 and go through 7777 + 7777.

Storage of Program and Data

All locations in core storage can be accessed at *random*. However, it is convenient to designate one area of memory as *program* area and another as *data* area. This allows for storing instructions in the *sequence* of execution and makes *automatic* sequential access possible. When we access memory for the purpose of removing a *data* word, we *supply* an address where the data is located. When we access memory for removing an *instruction*, we rely on the *control* section to give us the address of the next *sequential* instruction.

Review Exercises

1. What is a memory device?

2. Name the four methods of storage.

3. What are the two categories of computer memories?

4. Name two media used for mechanical storage.

5. What is the purpose of the card:
   a. Reader?
   b. Punch?

6. What is the Hollerith code?

7. In a five channel code for perforated paper tape, how many channels are available for each character?

8. What is the purpose of the small holes running lengthwise on a paper tape?

9. Flip flops, flip flop latches, and storage tubes are examples of devices used with which method of storage?

10. What is meant by nonvolatile storage?

11. List two major limitations of optical memories.

12. Name six storage media which utilize magnetic storage.

13. A certain method of recording on a magnetic surface represents one bits by a change in flux and zero bits by steady state flux. What is the name of this method?

14. Why are tapes unsatisfactory for central memories?

15. Describe access to information on magnetic disks to:
    a. A channel.
    b. Information within a channel.

16. What storage devices use magnetic heads for recording and retrieving information?

17. Why has thin film been unpopular as a storage method for central memories?

18. Ferrite cores have low reluctance and high retentivity. What bearing do these characteristics have on their suitability as storage devices?

19. Explain why a ferrite core must always contain either a binary zero or a binary one.

20. A computer has a word length of 32 bits (including both sign and parity bits). The words are vertically oriented in the coincident current core memory. The memory capacity is 4096 words.
    a. How many planes are used?
    b. How many cores are on each plane?
    c. How many flip flops are contained in the memory address register?
    d. How many flip flops are in the memory information register?
    e. How many cores are threaded by each X and Y line?
    f. A given X and a given Y line coincide through how many cores? What planes are these cores located on?
    g. How many sense amplifiers are used?
    h. How many inhibit drivers are used?
    i. How many cores are threaded by each sense and inhibit line?

21. Describe the action that occurs in a core when a zero is recorded.

22. How does a sense amplifier detect the presence of a:
    a. One?
    b. Zero?

23. Describe what happens to cores with ones and cores with zeros as a word is read from core memory.

24. Name the two types of memory cycles.

25. Each memory cycle is divided into two parts. What do we call these parts?

26. What are the inclusive octal addresses for a 64^2 memory?

27. How many X and how many Y lines are required for a 4096 word coincidence current core memory?

28. We have a ferrite core central memory which operates on odd parity.
   a. Explain the procedure for assigning parity and checking parity.
   b. Where does this take place?

29. Using the Hollerith code and reading from left to right, what characters are recorded on the card cutaway in Figure 9-21?

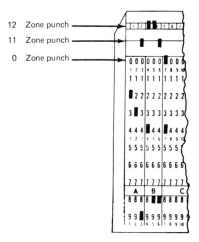

Fig. 9-21. Read the Characters.

central

processor

When we speak of the central processor, or central computer, we are grouping the major part of our computer system into a logical unit. This logical unit contains the central memory, the arithmetic section, and most of the control section. Every action that is initiated by a program instruction involves these three parts of the central processor. This chapter will analyze the interaction of these sections, while developing the functions of the control and arithmetic sections.

10-1 CONFIGURATION

The *three* sections of our central processor may be housed in a single table-top cabinet, or they may occupy several large cabinets. This largely depends upon the complexity and capability of the particular system. Whether they come in one cabinet or many cabinets has no bearing on the logical configuration. In any case, we have three logical sections: memory, arithmetic, and control, as shown in Figure 10-1.

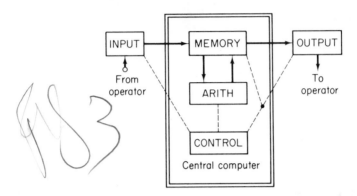

Fig. 10-1. Basic Computer System.

These three sections are designed to perform their individual functions in such a manner that they harmonize and complement one another. The *memory* section provides random access for storage of both program and data. The *control* section removes programmed instructions from memory, decodes them, prompts the arithmetic section to execute the instructions, and obtains the required data to go with each instruction.

The operator may insert the complete program and initial data through some appropriate input device. He can then manually set the address of the first instruction into the control section. The start button will now initiate automatic operation. The instructions of the program will be executed according to the programmer's prearranged order. The program is repetitive, which enables the automatic operation to continue for as long as desired.

Manual Controls

The central processor has many controls and indicators which make it possible for it to operate completely isolated from both input and output sections. Program and data can be entered into memory by manipulating manual push buttons and toggle switches. Automatic action can be established or the operator can manually step through small segments of the program.

Indicator lights constantly show the current contents of several registers and counters. A lighted indicator shows a set flip flop (a one condition); an unlighted indicator shows a cleared flip flop (a zero condition). The indicator panel generally has a light for each flip flop in the program address counter, memory address register, operation register, memory information register, accumulator, and others.

The lights are flashing on and off too rapidly for the eye to follow when they are in automatic operation. However, the action can be *halted* completely or *stepped* manually when the operator wishes to read the digital information from the indicators.

Key Components

We need to know the function and *interaction* of certain key components in order to follow program controlled arithmetic actions. We are using arithmetic actions in the broad sense to mean any action which is performed in or by the arithmetic section. The four basic mathematical operations are only a few of the arithmetic functions. The essential components are illustrated in Figure 10-2.

In chapter two we used this same block diagram with only slightly less detail. At that time, we also discussed *four* types of timing cycles. We had a *program cycle*, an *operate cycle A*, an *operate cycle B*, and a *memory cycle*. You should recall that the program cycle is used to obtain an *instruction*

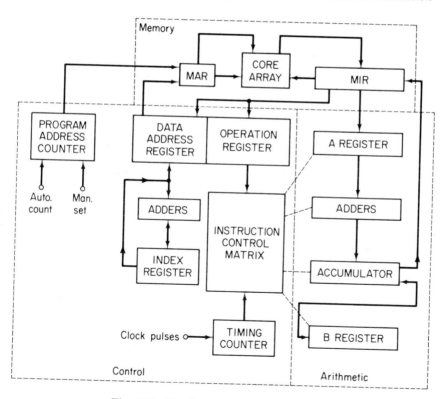

Fig. 10-2. Key Interacting Components.

from memory and *decode* it. The operate cycle A is used to obtain *data* from memory and perform some operation. Operate cycle B is used to clear a register in the central memory and *store* a new data word into it. Each of these three cycles requires a trip into memory. This means that we must have a memory cycle in support of *each* of our other cycles.

10-2   INTERACTION

We will bring our cycles back into play to emphasize the interaction of our essential components.

Program Cycle

The *address* of the instruction is provided by the program address counter. This address loads into MAR and a memory cycle begins. The *instruction* is read out into MIR and a copy is *rewritten* into the same core location. From MIR, part of the instruction word goes into the *operation* register and part into the *data address* register. The instruction word consists of a coded instruction (operation portion) and the address of the necessary data (address portion).

The *operation* register holds the instruction until the operation is completed. The *instruction control matrix* decodes the instruction by detecting the flip flop conditions. Levels from the detection operation combine with timing pulses to create signals which we call *commands*. Commands are timed pulses which control the details of the programmed action.

Operate Cycle A

The *data address* portion of the instruction word is sent back into memory to the MAR. Another memory cycle is initiated, the requested data are brought out to MIR, and a copy is replaced into core storage.

The instruction has already been decoded and action started. The A register has been cleared, and the data moves into the A register. The next action is determined by the nature of the instruction. We may move the data into a cleared accumulator, add it to the contents of the accumulator, perform any mathematical function, or make various decisions. In most cases, the *result* of the operation will end up in the *accumulator*.

Operate Cycle B

An operate cycle B may follow an A cycle under certain circumstances, but it is generally *preceded* by another program cycle in which a *store* instruction is decoded. At any rate, a data address passes from the data address register to MAR. A memory cycle is initiated which reads out the selected core contents. The transfer lines between core and MIR are *open* (disconnected) at this time, and MIR *does not* receive the information. The core register and MIR are *both* clear. The contents of the accumulator are moved into MIR, and from there, the memory cycle automatically transfers the bits into the core register.

Conventions

Just as we have different word lengths for different computers, we also have many other conventions which vary according to individual design. We have *single* address machines which require only *one* data address per instruction word. We have *multiple* address machines which use *two or more* addresses per instruction word. Some computers have instructions which each contain a number of bits exactly equal to a full computer word; these are computers with *fixed length* instructions. Other computers use *variable* length instructions, where a computer word may contain a part of an instruction, a whole instruction, or several instructions. Some computers move data bits in groups which are exactly equal to a computer word. When storing in memory, removing from memory, or moving from one register to another, each movement involves exactly one computer word. Other computers handle data in *syllables*. A syllable is a portion of a computer word and it usually contains about 12 bits. Another machine moves data in convenient batches which we call *bytes*. A byte is a basic unit of information usually consisting of eight binary bits. Bytes are also referred to as *characters*. Several consecutive bytes are used to form a computer word.

How numbers are represented in a computer forms another variable that is open to the designer's choice. It is fairly standard for numbers to be preceded by a *sign* bit, with a *zero* sign indicating a *positive* quantity and a *one* sign showing the quantity to be *negative*. The number representing the quantity is far less standard. Some computers are considered *fractional* machines. These have a binary point to the left of the most significant digit. The largest number that a fractional machine can contain is slightly *less than* one. Other computers consider the quantity to be a *whole* number.

Some computers use negative numbers in their *true* form while others have all negative numbers existing in the ones *complement* form. When a mathematical operation generates a negative number, the complement machine performs a ones complement on the number before sending it into storage.

With all these variable conventions, we could write a whole book without making a single concrete statement. It has been said that straightforward, definite statements concerning computers are impossible to make. We will be able to think in a less abstract fashion by establishing the characteristics of our *sample* computer.

Characteristics of Sample Computer

Our sample computer is a *stored* program, *single* address machine. It moves information in quantities equal to a *computer word*. The computer word consists of 16 bits; a *sign* and 15 magnitude bits. These words have an extra bit, an *odd parity* bit, when in the central memory.

We consider our quantities to be *whole* numbers, and all negative quantities are in the *ones complement* form. Sign bits are conventional; one for negative and zero for positive.

We have the *four* cycles previously mentioned. Each cycle is *six* μs in duration, and is divided into 12 timing intervals of 0.5 μs each. The clock pulses and timing pulses have a pulse recurrence frequency of 2 MHz.

This is the computer we have used in our block diagrams, and it is the sample computer we will continue to use. When discussing operations which are not applicable to this machine, we will try to make a distinction. These characteristics give us facts enough to establish timing charts, sample instructions, and other functional operations. At the same time, they describe a computer system of reasonable speed and sophistication, one which is fairly representative of the popular systems on the market.

10-3 CONTROL SECTION

We have already encountered some of the functions of the control section in conjunction with the central memory. The portion which appears on the block diagram of Figure 10-2 performs still more timing and control functions.

Timing

The timing pulses (12 per cycle) are furnished by a *ring counter* in our sample computer and in the majority of real systems. The timing pulses are of the same frequency as the clock pulses. In fact, they are clock pulses that are gated into action at critical points during the cycles.

Program Address Counter

This is a normal up counter which automatically prepares the address of the next sequential instruction. The operation is very simple. The operator manually sets the address of the *first* instruction into the counter. As soon as that instruction is brought from memory, one timing pulse is applied to the counter. This pulse *increases* the counter content by one, which is the address of the *second* instruction in the program.

Suppose that our program is stored into memory locations 500 through 550 (octal addresses). We *start* the program by setting 500_8 ($101\ 000\ 000_2$) into the program address counter and pushing the start button. The binary number $101\ 000\ 000$ is loaded into the MAR, and the memory cycle brings the instruction from location 500_8. Before this program cycle ends, we gate *one* timing pulse into the counter. The count *increases* to $101\ 000\ 001_2$ or 501_8 which is the address of our *second* instruction. The next program cycle uses this address to obtain the *second* instruction. Again, we step the counter to $101\ 000\ 010_2$ or 502_8 which is the address of the *third* instruction.

Situations arise which make it desirable to *skip* one, two, or three instructions. They are usually program generated. An instruction will check for some specified condition. If the condition is present, we skip the *next*, the *next two*, or the next three instructions. If the condition *is not* present we execute the next instruction in sequence. When one instruction is to be skipped, the detected condition gates an *extra* timing pulse into the counter. The *automatic* count increases the count by one and obtains the address of the next instruction in sequence. The *extra* count changes this address before it can be used. One instruction is skipped and a new sequence is started. When we desire to skip *two* instructions, the detected condition gates *two extra* pulses into the counter. Two instructions are skipped and a new sequence is started. When three is added to the count, the next three instructions are skipped.

Data Address Register

This register provides temporary storage for the *data address* portion of each instruction word. This is the address of the *operand*, the data needed to perform the instruction. For most instructions, the address is simply held here until the program cycle ends; then it loads into MAR.

There are situations which make it necessary to *modify* the data address. At a certain point in the program, it may be necessary to repeatedly perform a given instruction with different data each time. One way to obtain access to different data locations is to *calculate* the data address. The index register and the address adder enable such calculations.

When an addition is performed in this area, the contents of the index register are added to the contents of the address register, and the *results* become the new data address in the address register.

A number can be placed into the index register by certain program instructions. The number can then be incremented or decremented each time it is used. This provides access to many data locations by calculating the data address.

In our block diagram, we have shown only *one* index register. In actual practice, we usually have from *five to twenty*. Some systems have the ability to use *any* central memory location as an index register. Many of the decision making instructions use the contents of specified index registers as a basis for making decisions.

One type of instruction directs a *jump* of instruction control to a different area of the program. For example, at one point in the program, it may be necessary to obtain a block of data which is stored on tape or in some other *auxiliary* memory. At that point in the program, we insert a JUMP instruction (sometimes called a BRANCH instruction). This instruction *moves* control to a section of the program which will select the appropriate input device and move the required data into central memory. The operation portion of the instruction directs the control section to take the next instruction from the location indicated by this instruction's data address.

When the control unit decodes such a JUMP instruction, it clears the program address counter and moves the contents of the data address register into this counter. The next instruction is brought from this *specified* location and the program address counter starts its *new* program sequence from that address.

When the job is done, *another* jump is required to transfer control back to the point where we interrupted our main program.

Small programs which are set apart from the main program to accomplish such a specialized task are called *subroutines*. The instructions directing control to a subroutine are sometimes called *subroutine jumps*. The same instruction with a different address usually brings us back to the main program when the subroutine is completed. The whole class of instructions is sometimes called subroutine *jump* and *return*.

Decoding Instructions

The *instruction control matrix* works similarly to any other decoder. The *operation* portion of each instruction word establishes a certain configuration in the flip flops of the operation register. The decoder *combines* the levels from these flip flops with timing pulses to generate the necessary *commands* and signals to execute the indicated action.

The list of computer instructions can be broken down into only a few distinct *classes*. A group of instructions which has many actions in common is called a class. For instance, all instructions which perform mathematical operations (in most computers) are variations on the ADD instruction. For this reason, the *add class* of instructions includes clear and add (CAD), add (ADD), and subtract (SUB).

Another class of instructions moves data between central memory and terminal equipment. This is the *input/output* class. The group of instructions which shifts data is called the *shift* class.

Let's take a few instructions and develop timing and control commands. This can be done with the logic shown in Figure 10-3.

Coming into the upper left of the diagram we have five lines. They are labeled CAD, ADD, SUB, MUL, and DIV. One of these lines will carry a high level from the operation register when the indicated instruction is present. This level will *remain high* throughout the operate A cycle while the instruction is being executed.

Some of the abbreviations on the diagram may need explaining.

MAR = Memory address register
MIR = Memory information register
ADR = Address register
REG = Register
CNTR = Counter
XFER = Transfer
ACC = Accumulator
C(—) = Contents of item identified in the parentheses.
SL = Shift left

There is very little difference in the actual operation of the CAD, ADD and SUB instructions. Signals which perform key functions are called *commands* and are sometimes identified by command numbers. To the right of the AND gates, we have numbers in *parentheses* from 1 to 17. These are numbered commands. The task performed by each command is indicated on the output line.

CAD

CAD·TPO = (1) to clear the MAR
CAD·TPI = (2) to clear MIR, XFER C(ADR) to MAR, start memory
            cycle and clear the A register.
CAD·TP4 = (3) to check parity and XFER C(MIR) to A register.
CAD·TP4 (through AND 6) = (6) to clear ACC.
CAD·TP6 (through AND 11) = (11) ADD pulse to start addition.

ADD

Notice that this instruction generates exactly the same commands as we have just seen for CAD with one exception. The exception: AND 6 is not activated, command 6 is not generated, and the accumulator is not cleared.

By clearing the accumulator before adding during CAD, we add a word from memory to C(ACC), which is all zeros. This effectively moves the C

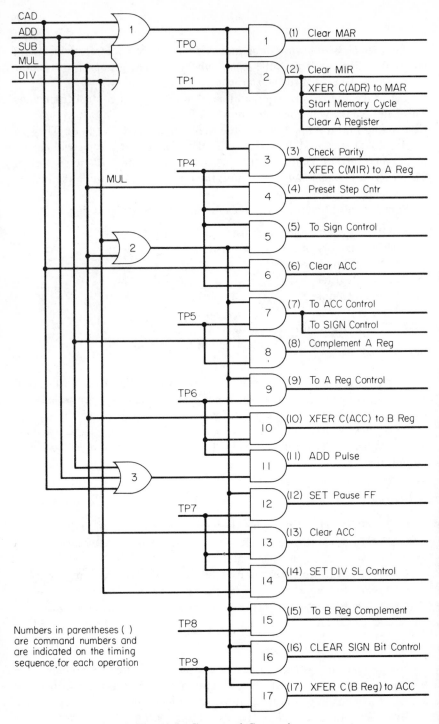

Numbers in parentheses ( ) are command numbers and are indicated on the timing sequence for each operation

Fig. 10-3. Command Generation.

(memory location) into the accumulator. With the ADD instruction, our objective is to take the sum of the C(memory location) and C(ACC) and place this sum in the accumulator. The difference is command 6 at TP4.

SUB

SUB·TP0 = (1)
SUB·TP1 = (2)
SUB·TP4 = (3)
SUB·TP5 = (8)
SUB·TP6 = (11)

Notice that the only difference between ADD and SUB is that the subtract generates command 8 at TP5. This command *complements* the contents of the A register just before starting the addition.

Pause Control

The multiply and divide instructions are a bit more complicated. The primary reason for this is the multiple operations which require *more time* than we have in a normal timing cycle. This makes it necessary to *stop* the timing pulse ring counter after the instruction is initiated. When the multiple operations are completed, we again *start* the ring counter.

This interruption of normal timing is called an arithmetic *pause*, or simply a pause. The pause is initiated at TP7, and when it ends, the next timing pulse will be TP8 of the same timing cycle. The ring counter simply picks up where it left off and continues its normal count as if nothing happened.

During the pause, timing from another counter provides pulses to perform the repeated additions or subtractions. This service is generally performed by a multiply and divide ring counter.

MUL

MUL·TP0 = (1)
MUL·TP1 = (2)
MUL·TP4 (through AND 3) = (3)
MUL·TP4 (through AND 4) = (4)
MUL·TP4 (through AND 5) = (5)
MUL·TP5 (through AND 7) = (7)
MUL·TP5 (through AND 8) = (8)
MUL·TP6 (through AND 9) = (9)
MUL·TP6 (through AND 10) = (10)
MUL·TP7 (through AND 12) = (12)
MUL·TP7 (through AND 13) = (13)

What have we accomplished thus far? At TP0 command 1 cleared MAR. At TP1, command 2 cleared MIR, transferred the data address into MAR, started the operate A memory cycle, and cleared the A register. At TP4 we checked parity on the data that came from core into MIR and then transferred this data into the A register. Also at TP4, we preset a count into the step counter to control the number of repetitive additions that will be performed. If our accumulator contains a negative number, command 5 at TP4 complemented the sign control flip flop. If the A register has a negative number, at TP5 command 7 complemented the sign control flip flop. (If both numbers were positive commands, 5 and 7 did nothing). At TP6 command 9 complemented the C(A REG) if that quantity was negative. Also at TP6, command 10 moved C(ACC) into the B register. At TP7 we generated commands 12 and 13. Command 12 set the *pause* flip flop to stop our timing pulse counter, and command 13 cleared the accumulator.

All of this is setting the stage for the performance of a multiplication. We are now in a *pause* and our repetitive operations are under the control of a special counter. This counter will cause an addition on each clock pulse until the count in our preset step counter is reduced to a zero.

A *zero* in the step counter stops the operation, and *clears* the pause flip flop. When the pause flip flop is cleared, this *restarts* our timing pulse counter. The next pulse is TP8.

$$MUL \cdot TP8 = (15)$$
$$MUL \cdot TP9 = (16)$$
$$MUL \cdot TP9 = (17)$$

What did these three commands accomplish? At TP8 the most significant bits of the result of the multiplication were in the B register. These three commands assigned the proper *sign* bit to that quantity, *complemented* it if it was negative, and *shifted* these bits into the accumulator. This is an end a-round shift which *exchanges* the contents of the accumulator and B register.

DIV

$$DIV \cdot TP0 = (1)$$
$$DIV \cdot TP1 = (2)$$
DIV·TP4 (through AND 3) = (3)
DIV·TP4 (through AND 5) = (5)
DIV·TP5 (through AND 7) = (7)
DIV·TP5 (through AND 8) = (8)
DIV·TP6 (through AND 9) = (9)
DIV·TP7 (through AND 12) = (12)
DIV·TP7 (through AND 14) = (14)

These are the preliminary commands which start the action and move the timing up to the pause which starts at TP7. During the pause, the special counter controls the repetitive steps of subtraction.

When the division is completed, the pause flip flop will be cleared to restart the timing pulse ring counter.

$$DIV \cdot TP8 = (15)$$
$$DIV \cdot TP9 = (16)$$
$$DIV \cdot TP9 = (17)$$

These last three commands perform the same job as they did with multiply.

The multiplication was performed by adding and shifting right. The division was accomplished by subtracting and shifting left.

10-4 ARITHMETIC ACTION

All arithmetic actions are performed through step by step *prompting* from the control section. We have just seen how the control section generates its prompting commands. Our arithmetic section does a great deal more than simple arithmetic, but if we can perform addition, subtraction, multiplication, and division, we will need very little additional logic.

Simplified Arithmetic Logic

Nearly all of the actions described for the add class instructions can be performed with the logic in Figure 10-4. The seven bit computer word has been used to shorten the operations. It has little bearing on the actual functions.

Examine the logic in Figure 10-4 and acquaint yourself with the key registers, counters, and flip flops. Notice in the lower left a ring counter with outputs labeled MDTP0 through 4. This is the counter which controls our action during the arithmetic pause. The counter produces pulses at *five time* the clock frequency, giving us five timing pulses for each 2 MHz clock pulse. The MDTP is an abbreviation for multiply and divide timing pulse. A complete addition or subtraction is performed with each cycle of MDTP0, 1, 2, 3, and 4. The 2 MHz clock pulse resets the counter to zero to begin the next cycle.

The numbers in parentheses are *command* numbers that we generated previously. Now let's pick up the timing for each of the add class instructions, and analyze the logic actions.

Function of CAD

At TP0 we clear the MAR. At TP1 we transfer the data address into MAR, start the memory cycle, and clear the MIR. These last three tasks are

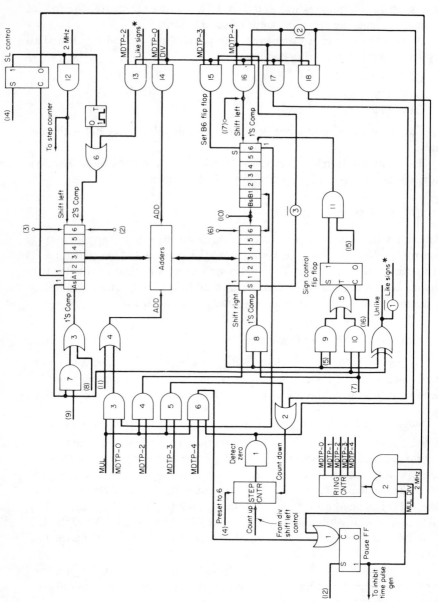

Fig. 10-4. Simplified Arithmetic Logic.

284

accomplished by command 2. On our arithmetic logic, we have command 2 being applied to the A register. This is still TP1, and we are clearing the A register to receive the data from the MIR.

CAD·TP4 generates command 3. This command causes a *parity check* of the data word which is now in the MIR. If we have odd parity, the same command moves the data from the MIR to the A register.

CAD·TP4 also generates command 6 which *clears* the accumulator.

CAD·TP6 generates command 11 which *starts* the addition. The adder logic takes the sum of the contents of the A register (the data word from memory) and the accumulator (which has been cleared to all zeros). The original contents of the accumulator are *replaced* by the sum of the two numbers.

Function of ADD

We have commands 1, 2, and 3 which perform the same functions as in CAD. This time one of our quantities is *already* in the accumulator; therefore, command 6 *is not* generated.

At TP6 we have one quantity in the *A register* (the data word from memory) and the other in the *accumulator*. The accumulator contents have been generated by the previous arithmetic operation. At this time we generate command 11, which causes the adder to *calculate* the sum of our two quantities. This sum replaces the original contents of the accumulator.

Function of SUB

Again we have commands 1, 2, and 3. These commands are *common* to all add class instructions and to nearly all other instructions. Their functions are always the same.

To subtract, we must *complement* the contents of the A register before we start the addition process. This is accomplished when SUB·TP5 generates command 8.

At TP6 we again have command 11 which causes a summation of the complemented number in the A register and the contents of the accumulator. The sum of these two numbers is the *difference* between the data word in memory and the original quantity in the accumulator. This result resides in the accumulator when the instruction terminates.

Function of MUL

After our common commands 1, 2, and 3, we generate commands 4 and 5 at TP4. Command 4 presets the step counter. The logic has a six bit word plus a sign bit; therefore, the step counter is set to a count of six. This count in any computer is set to a number which equals the number of magnitude bits in the computer word. Command 5 is applied to AND 9. If sign of the accumulator is negative, AND 9 is activated, which applies a signal through OR 5 to complement the sign control flip flop. This flip flop is normally clear, and the complement pulse set it.

MUL·TP5 generates command 7. If the accumulator sign is negative, AND 8 is conditioned, in which case, command 7 complements all bits in the accumulator. If the A register sign is negative, AND 10 is conditioned, which allows command 7 to pass through and recomplement the sign control flip flop.

Our situation is this: if both numbers are positive, the sign control flip flop has not changed; it is still clear. If both numbers are negative, the sign control flip flop has been complemented twice, from clear to set and back to clear. If either number is positive and the other negative, the sign control flip flop has been complemented only once, from clear to set. The object of this maneuver is to determine the proper sign for the product. The proper sign, of course, is positive if the multiplier and multiplicand have like signs and negative if their signs are different. The state of this flip flop after TP5 is the proper sign of our final product.

MUL·TP6 generate two commands, 9 and 10. Command 9 goes to AND 7, which is conditioned only if the A register sign is negative. If this is a negative number, command 9 complements all bits in the A register. Command 10 transfers the contents of the accumulator into the B register.

MUL·TP7 generate commands 12 and 13. Command 13 clears the accumulator, and command 12 sets the pause flip flop.

The level from the one side of the pause flip flop inhibits the timing pulse counter and conditions one leg of AND 2. A second leg of AND 2 is conditioned by the fact that our shift left control flip flop is in a clear condition. A third leg on AND 2 is conditioned at all times during either a multiply or divide instruction. These three legs are then conditioned as long as our pause flip flop is set. The 2 MHz clock pulse will pass through AND 2 and start the MDTP ring counter. The counter will produce pulses 0, 1, 2, 3, and 4; then the next clock pulse starts it over at pulse 0. This action will continue as long as the pause flip flop is set.

MUL·MDTP0·B6 activate AND 3 and start an addition. MUL·MDTP2 shift the accumulator contents to the right. MUL·MDTP3 activate AND 5 and reduce the count in the step counter by one. MDTP4 is blocked by AND 6 as long as the step counter contents are *other than* zero.

Immediately after MDTP4, the 2 MHz clock pulse through AND 2 restarts the counter at MDTP0.

With each MDTP2 the contents of the accumulator and the B register are shifted one place to the right. Each MDTP0 causes an addition, provided the *LSB of the B register is a one* at that time. MUL·MDTP0·$\overline{B6}$ inhibit the addition for that MDTP cycle. In other words, cycles which find B6 add and shift; cycles which find $\overline{\text{B6}}$ shift but *do not* add.

In the shifting operation, the B register sign bit does not change. The LSB of the accumulator moves into B1, while all magnitude bits of accumulator and B register move one place to the right. Check this path on the logic in Figure 10-4.

Since we preset the step counter to six, our logic will cause the MDTP ring counter to execute six cycles. At the close of the sixth cycle, AND 6 is conditioned by the zero count. This allows MDTP4 to pass through AND 6, through OR 1 and clear the pause flip flop.

Clearing the pause flip flop removes the inhibit level from our timing pulse counter, and at the same time, it removes the conditioning level from AND 2 which inhibits the MDTP counter. Control reverts to our timing pulse counter, which picks up the next timing pulse. This is TP8.

This *is not* a double precision machine; so the results of the multiplication now reside in the B register.

MUL·TP8 generate command 15 which is applied to AND 11. If our sign control flip flop is clear, AND 11 is inhibited, and command 15 does nothing. This is the proper condition when both original numbers had like signs. The sign bit of the B register will be clear at this time, and since our result is positive, we will leave it cleared. If the sign control flip flop is set, and it is if the two original numbers had different signs, AND 11 is conditioned, and command 15 passes through to complement all bits in the B register. This means that our product is negative, and we have just set the sign bit to a one and placed the magnitude bits in the ones complement condition.

At the next count, MUL·TP9 generate command 16 which is applied to the clear side of the sign control flip flop. This assures that the flip flop will be in the zero state the next time we want to use it. MUL·TP9 also generate command 17 which interchanges the contents of the accumulator and the B register. This is accomplished by executing a double, left, logical, end around shift. The number of shifts will be equal to the number of bits in a computer word; in this case seven shifts are accomplished.

It is not essential that command 17 be executed. Some machines are designed to leave the product where it is at the end of the multiplication process. This is unimportant. The programmer knows where it is, and he can accomplish shifting and rounding off by using another instruction.

Example of Multiply Function

The following example will prove the feasibility of using this type of logic for multiplication. We will take two numbers and multiply machine fashion. We'll perform the actions as dictated by the commands and timing pulses.

Choosing a relatively simple number, let's multiply a $+5$ in the accumulator by a —3 from a memory location. This will give us an octal product of —17. In binary form we have:

$$0.000101 \times 1.111100 = 1.110000$$

Keep in mind that negative numbers are in complement form. Do not be confused by the period between the sign and the MSB. This is the conventional method of separation. This period is not considered as a binary point except when dealing with a fractional machine.

A Register	*Accumulator*	*B Register*	*Action*
1.111100	0.000101	0.000000	(3) (4) SC = 6
0.000011	0.000101	0.000000	(9) COMP A
0.000011	0.000101	0.000101	(10) C(ACC) to B Reg
0.000011	0.000000	0.000101	*(13) Clear acc*
0.000011	0.000011	0.000101	0–1 ADD
0.000011	0.000001	0.100010	2 SHIFT RIGHT
0.000011	0.000001	0.100010	3 SC = 5
0.000011	0.000001	0.100010	*4 no action*
0.000011	0.000001	0.100010	0–1 No ADD
0.000011	0.000001	0.110001	2 SHIFT
0.000011	0.000000	0.110001	3 SC = 4
0.000011	0.000000	0.110001	*4 No action*
0.000011	0.000011	0.110001	0–1 ADD
0.000011	0.000001	0.111000	2 SHIFT
0.000011	0.000001	0.111000	3 SC = 3
0.000011	0.000001	0.111000	*4 no action*
0.000011	0.000001	0.111000	0–1 No ADD
0.000011	0.000000	0.111100	2 SHIFT
0.000011	0.000000	0.111100	3 SC = 2
0.000011	0.000000	0.111100	*4 no action*
0.000011	0.000000	0.111100	0–1 No ADD
0.000011	0.000000	0.011110	2 SHIFT
0.000011	0.000000	0.011110	3 SC = 1
0.000011	0.000000	0.011110	*4 no action*
0.000011	0.000000	0.011110	0–1 No ADD
0.000011	0.000000	0.001111	2 SHIFT
0.000011	0.000000	0.001111	3 SC = 0
0.000011	0.000000	0.001111	*4 clear pause FF*
	0.000000	1.110000	(15) COMP B
	[1.110000]	0.000000	(17) C(B) TO ACC

Function of DIV

Commands 1, 2, 3, 5, 7, 8, 9, and 12 perform the same action as in multiply.

DIV·TP7 generate command 14 which sets the shift left control flip flop. Instead of presetting the step counter to a specific number, we use the C(A register) to determine the number of steps. When command 14 sets the shift left control flip flop, AND 12 is conditioned. As long as this flip flop is set, each 2 MHz clock pulse will pass through AND 12, add one to the step counter, and shift C(A register) one position to the left. This shift will contine until the MSB of the A register becomes a one. When this happens, AND 12 is inhibited, shifting is stopped, and the step counter becomes a down counter with a specific count already set. The count in the counter is equal to the number of shifts required to move a one bit into the MSB position of the A register.

The high level from the zero side of the shift left flip flop enables AND 2. The 2 MHz clock pulses will now start the MDTP counter, and it will cycle as described for the multiply instruction. When we set the shift left control FF, the level sets a one shot multivibrator. After a short delay, which allows time for shifting, the one shot reverts to its normal clear position. The high from the zero side of the one shot goes through OR 6 and performs a twos complement of the A register contents.

We use the twos complement during divide because it speeds up the action by eliminating the need for end carry.

Notice that the sign flip flops of A register and accumulator are monitored by an exclusive OR gate. The output of the exclusive OR is low at any time that the two signs are the same. This low is inverted to enable AND 13. DIV·LIKE SIGNS·MDTP2 activate AND 13 and send a pulse through OR 6 to recomplement (twos complement) the contents of the A register.

DIV·MDTP0 activate AND 14 to cause an addition. MDTP1 causes no action but the time is needed to complete the addition. MDTP2 complements contents of A register when the sign matches the sign in the accumulator. MDTP3 sets the B6 flip flop through AND 15 when the accumulator sign is a zero. DIV·MDTP4·STEP COUNT $\bar{0}$ activate AND 16 and shift combined contents of B register and accumulator one position to the left.

This concludes the first MDTP cycle. We will complete one cycle for each count in the step counter plus one additional cycle. When the step counter reaches zero, AND 18 is enabled while AND 16 is disabled. The MDTP4 of the next cycle passes through AND 18 to OR 1, through OR 1, and clears the pause flip flop. The timing pulse counter picks up on TP8.

DIV·TP8 generate command 15 to complement the contents of the B register if the sign control flip flop is set. This flip flop is set only when our original numbers were of different signs. This means that our quotient is a negative number which must be placed in the complement form.

DIV·TP9 generate commands 16 and 17. Command 16 clears the sign control flip flop, and command 17 exchanges the contents of the accumulator and B register. This concludes the divide instruction and leaves the quotient in the accumulator.

It should be kept in mind that this limited logic has no provision for calculating or preserving the remainder. How this is handled is the designer's choice but he usually provides for retention of the remainder until it can be checked. If it is large enough to be significant the quotient is usually increased by one; otherwise it is dropped.

Example of Divide Function

Let's divide $+31_8$ by $+5$. This should work out as follows:

$$0.011001/0.000101 = 0.000101$$

A Register	Accumulator	B Register	Action
0.000101	0.011001	0.000000	
0.101000	0.011001	0.000000	SHIFT AND SET SC = 3
1.011000	0.011001	0.000000	COMP A (2s COMP)
			MDTP
1.011000	1.110001	0.000000	0.1 ADD
0.101000	1.110001	0.000000	2 COMP A
0.101000	1.110001	0.000000	3 0 to B6
0.101000	1.100010	0.000000	*4 shift SC = 2*
0.101000	0.001010	0.000000	0.1 ADD
1.011000	0.001010	0.000000	2 COMP A
1.011000	0.001010	0.000001	3 1 to B6
1.011000	0.010100	0.000010	*4 shift SC = 1*
1.011000	1.101100	0.000010	0.1 ADD
0.101000	1.101100	0.000010	2 COMP A
0.101000	1.101100	0.000010	3 0 to B6
0.101000	1.101100	0.000100	*4 shift SC = 0*
0.101000	0.010100	0.000100	0.1 ADD
1.011000	0.010100	0.000100	2 COMP A
1.011000	0.010100	0.000101	3 1 to B6
1.011000	0.010100	[0.000101]	4 No SHIFT—CLEAR
			PAUSE FF

This action stops with the fourth MDTP-4. The answer is still in the same

B register position because we had no MDTP4 through AND 16 to execute the left shift.

Let's take another example. This time we will divide +24 by +12. Binarily this should be:

$$0.010100/0.001010 = 0.000010.$$

A Register	Accumulator	B Register	Action
0.001010	0.010100	0.000000	
0.101000	0.010100	0.000000	SHIFT AND SET SC = 2
1.011000	0.010100	0.000000	COMP A (2s COMP)
			MDTP
1.011000	1.101100	0.000000	0–1 ADD
0.101000	1.101100	0.000000	2 COMP A
0.101000	1.101100	0.000000	3 0 to B6
0.101000	1.011000	0.000000	*4 shift SC-1*
0.101000	0.000000	0.000000	0–1 ADD
1.011000	0.000000	0.000000	2 COMP A
1.011000	0.000000	0.000001	3 1 to B6
1.011000	0.000000	0.000010	*4 shift SC = 0*
1.011000	1.011000	0.000010	0–1 ADD
0.101000	1.011000	0.000010	2 COMP A
0.101000	1.011000	[0.000010]	3 0 to B6
			4 NO SHIFT—CLEAR
			PAUSE FF

It should be noted that this limited logic cannot accurately divide all numbers. For instance, if we had divided 24 by 2 in the last example, we would have obtained a quotient of 13 instead of 12. When logic diagrams are simplified for the sake of explanation, inaccuracies of this type must be expected.

10-5 OVERFLOW

When we perform arithmetic on paper, we have no problem with the number of available columns. When we exceed the limits of one column we simply carry information into a higher order column. The computer lacks this flexibility. It has one column for each bit position in the computer word. A four bit word indicates four columns available for arithmetic processes. An 80 bit word indicates 80 columns, but it is still limited.

The largest number that a register can contain is called the modulas of that register. When an arithmetic operation creates a number which is larger

than the modulas of the accumulator we lose data because our most significant column is outside the limits of our register. We refer to this condition
as overflow.

Overflow Conditions

It is possible to generate results too large for the machine when adding
numbers of like signs and when subtracting numbers of unlike signs. The sign
bit is involved in the addition process the same as all other bits. One overflow
condition exists when addition of two positive numbers produces a negative
result. Another overflow condition is present when addition of two negative
numbers produces a positive result. The third condition is actually a variation
of the first two since subtraction is a process of complementing and adding.

End carry is *not* necessarily overflow. Let's examine a few operations to
clarify this point. We'll use a four bit word.

1. Add $+5$ and -4.

$$
\begin{array}{r}
0.101 \\
1.011 \\
\hline
0.000 \\
1 \\
\end{array}
$$

$$
\underset{\overline{0.001}}{\text{END CARRY}}
$$

In this case, the positive number was the larger number and the sum
should be a positive number. When we add these positive and negative numbers, we are actually taking the difference between $+5$ and -4. The end result
is a $+1$ which we obtained after the end carry was added to our first result.
This type of end carry *is not* overflow.

2. Add $+4$ and $+6$.

$$
\begin{array}{r}
0.100 \\
0.110 \\
\hline
1.010 \\
\end{array}
$$

We had no end carry, but the carry from the MSB position changed our
sign bit from positive to negative. The result is obviously in error. This is
overflow.

3. Add -6 and -4.

$$
\begin{array}{r}
1.001 \\
1.011 \\
\hline
0.100 \\
\end{array}
$$

$$
\text{END CARRY} \longrightarrow \underset{\overline{0.101}}{1}
$$

Here we had end carry but *did not* have a carry from the MSB. The sum of two negative numbers appears to be a positive quantity. This is an incorrect operation; it is overflow.

Detection of Overflow

Since it is possible to create inaccurate data through the arithmetic process, our machine must be capable of detecting the overflow condition. This can be done with the logic represented in Figure 10-5.

When two *positive* numbers are being added, the sign bit control is *deactivated*. If a carry is generated from the MSB adder (A1 & B1), it will activate the overflow logic.

When two negative numbers are being added, the sign bit control is activated. This enables the end carry logic as well as the line to the overflow logic. Now a carry from the MSB adder complements the overflow flip flop to the set state. This is all right provided we also have end carry. The end carry will feed through the sign control logic and complement the overflow flip flop back to the clear state. In other words, the end carry *cancels* the overflow indication before the information is acted on. However, a carry from the MSB *without* end carry or end carry without a carry from the MSB will leave the overflow flip flop in a set condition. When the operation is over, the overflow will be checked. If this check pulse finds the overflow flip flop set, it activates the AND gate and initiates the overflow alarm.

Controlled Overflow

Overflow from an arithmetic operation *is not* necessarily a bad thing. Many operations will cause overflow intentionally. When overflow is planned, the alarm logic can be inhibited, and computer operation will continue. Unplanned overflow will light an indicator on the operator's console and, in many systems, it will cause an audible alarm. The operator can then make the decision to stop and correct the data or cancel the alarm and continue.

10-6 SHIFT FUNCTION

We have discussed the various types of shifts, and examined the operation on a block diagram. Now we will examine the logic for accomplishing a shifting action.

Right Shift Logic

Logic for shifting action is extremely simple, as you can see in Figure 10-6. With a simple gating arrangement such as this we can move the entire register contents one position to the right with each shift pulse.

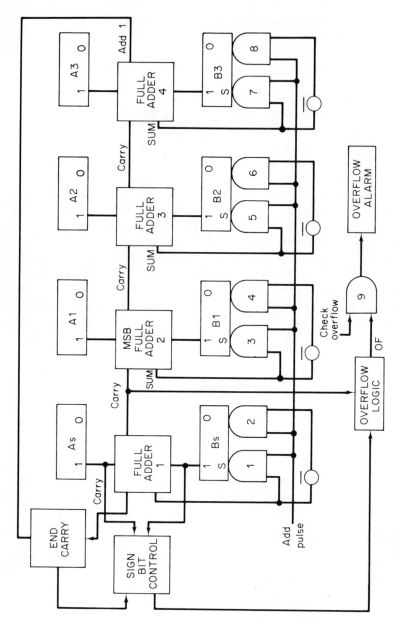

Fig. 10-5. Adder with Overflow Detection.

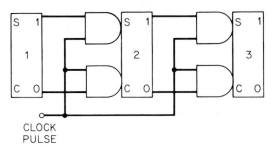

Fig. 10-6. Right Shift.

Many shifting actions are *automatic* because they are built into the hardware. Other shifts are software functions and must be *programmed*. When programming shifts, the shift instruction specifies what type of shift and how many positions. The desired number of shifts is loaded into the up-down step counter. The shift pulse is gated into the shift logic by an AND gate which is activated by the step counter being $\bar{0}$. Each shift pulse decrements the count by one at the same time that it moves the register contents by one position.

When the step counter reaches a zero count, it inhibits the shift pulse and stops the operation.

Left Shift Logic

Left shift logic is the same as right shift in reverse order. The only difference is the relative position of the flip flops which feed the gates. This is illustrated in Figure 10-7.

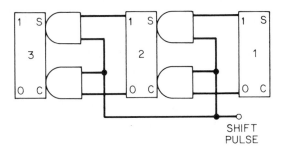

Fig. 10-7. Left Shift Logic.

Scaling Numbers by Shifting

When an engineer is using a constant of 10,000 in mathematical computations, he generally *scales* the number to a more convenient figure. Chances are that he will scale it to 10 and multiply the results by 1,000 or he will scale it to 1 and multiply the results by 10,000.

The computer *programmer* also finds scaling a useful tool, and he can accomplish this by shifting the number. In binary notation, shifting one place to the left multiplies the number by two; one right shift divides the number by two. If we allow n to equal the number of shifts and X to equal the original number, we can write formulas for our scaling operations:

$$\text{Left shift:} \quad X \cdot 2^n$$
$$\text{Right shift:} \quad X \cdot 2^{-n}$$

Example: $0.001\ 111_2$

$$1\ \text{SHIFT LEFT:} \quad 0.011\ 110_2$$
$$2\ \text{SHIFTS LEFT:} \quad 0.111\ 100_2$$
$$\text{ORIGINAL NUMBER:} \quad 0.001\ 111_2$$
$$1\ \text{SHIFT RIGHT:} \quad 0.000\ 111_2$$
$$2\ \text{SHIFTS RIGHT:} \quad 0.000\ 011_2$$

It is up to the programmer to keep track of his scale factors. The processor will operate with any numbers it has and produce results accordingly.

Review Exercises

1. What portion of a computer system is incorporated into the central computer?

2. Draw the basic system block diagram and indicate the blocks that are in the central computer.

3. To what extent is the central computer automatic?

4. Describe the part played by the control section when the central computer is under program control.

5. After storing both program and data, what action must be taken by the operator in order to establish automatic operations?

6. After the first instruction of a program has been obtained from memory, how are addresses of subsequent instructions obtained?

7. What provisions are made for manually storing and monitoring information?

8. Figure 10-8 illustrates the MIR with indicator lights. Which indicators are lighted when the register contains the indicated binary configuration?

9. The word in Figure 10-8 is on its way to core storage. At what point in the machine was the parity bit set?

10. Assuming that the word in Figure 10-8 is a valid data word, what type of parity does the machine use?

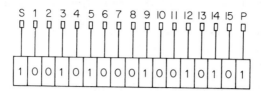

Fig. 10-8. MIR with Indicators.

11. Name the four types of cycles.

12. List the cycles necessary to decode and execute each of these three instructions:
   a. CAD.
   b. ADD.
   c. STR (STORE).

13. List four ways of changing the address in the program address counter.

14. In step 21 of our main program we will need a square root table which must be calculated. Describe the programmed actions from the beginning of step 20 to the start of step 21.

15. What is the common term for the action you described in item 14?

16. We have decoded an instruction with a data address of 775_8. If we modify this address with the contents of an index register which contains 3, we will obtain data from which core location?

17. Describe the path of transfer and the key actions during the execution of the ADD instruction.

18. Assuming that negative numbers are in ones complement form, what is the octal magnitude of the number in Figure 10-8?

19. Using the logic diagram in Figure 10-3, write the Boolean equation for command:
   a. 5.
   b. 7.
   c. 11.

20. AND 6 on the logic diagram of Figure 10-3 is inoperative. How does this affect CAD instruction?

21. AND 8 of Figure 10-3 is inoperative. Which instruction is affected and in what manner?

22. Why must a pause be used in the MUL and DIV instructions?

23. What combination of conditions will terminate a pause?

24. Why do we use the twos complement during a DIV instruction?

25. On the logic diagram in Figure 10-4, what is accomplished by commands:
    a. 5?
    b. 7?
    c. 15?

26. During the execution of a DIV instruction, how often do we execute the twos complement on the contents of the A register?

27. What is the frequency of the MD timing pulses?

28. How much time is required to complete each MDTP cycle?

29. During the multiplication process, what conditions will cause an addition?

30. In a MUL MDTP cycle which does not add, what inhibits the addition?

31. When preparing to divide, what determines the count to be set in the step counter?

32. A divide instruction sets a count of 6 into the step counter. How many MDTP cycles will be executed?

33. During DIV, what conditions cause:
    a. B6 to be set?
    b. B6 to be left clear?

34. Describe the condition called overflow.

35. Which of these additions will cause overflow:

    a. 0.111 010   b. 1.111 010   c. 1.000 000
        0.010 101   1.010 101   1.000 001

input and output

We sometimes have a tendency to think of the input and output sections of our computer system as a *collection* of miscellaneous peripheral *devices*. These sections are largely composed of devices to be sure, but they are a great deal more. Actually these two sections are the *links* between the central computer and the outside world. The input section enables the operator to *send* directions and data to the central computer. The output section enables the central computer to *return* information to the operator.

The input section *encodes* masses of operator generated information and furnishes it to the central computer in the computer's own language. The output section *decodes* masses of computer generated information and furnishes it to the operator in his own language. For instance, a keyboard operator may type a coded message such as "M4-2". The encoding circuits change this to binary digits which tell the computer to run a diagnostic program. The computer returns a group of binary bits to the keyboard decoder circuits.

299

When decoded, the proper keys will be activated to print the results of the diagnostic test in plain English.

11-1 TIMING AND CONTROL

The combined input and output transactions represent the *majority* of the operational time for any system. The control program, and all initial data, must be entered bit by bit. New programs and subroutines must be obtained from auxiliary memories from time to time. New data is constantly arriving from the operator, storage devices, and communications lines. Output information is moved to tapes, ~~drums~~, displays, disk storage, keyboards, printers, and tape storage.

The *fastest* machinery in our system is the central computer, which performs complete operations in a matter of a few ns, or at the very worst, a few μs. The *slowest* functional unit in the system is the human operator. The other devices run the gauntlet between these two extremes of speed and efficiency.

Timing and control in the input and output sections are major engineering problems. The problems are similar to those we would encounter in trying to establish intelligent contact between an ox cart and a space craft moving at the speed of light. Each device functions at its own *optimum* speed from very slow to very fast, but to the central computer, all of them are very slow.

The central computer must lose a lot of time in order to deal with these devices. The most efficient system is the one which handles the inputs and outputs (I/O) with the least waiting time. There are *two* basic systems of timing and control for I/O operations. We call them the *synchronous* and the *asynchronous* systems.

Synchronous System

The *completely* synchronous system has all details of every operation under the *supervision* of the central computer. For instance, an instruction is decoded which says, "Read 2000 words from the card reader." All computations are *suspended* while the cards are read and the 2000 words are moved into memory one at a time. *No other* instructions are decoded during the

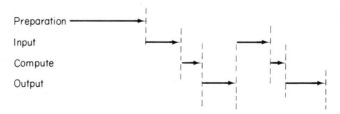

Fig. 11-1. Actions in a Synchronous System.

operation because the I/O instruction must remain in the operation register and generate the control commands.

This system is slow, and it *wastes* the majority of the central processor's operating time. Actual productive time is represented by *short* intervals of computation interspersed with long intervals of I/O operations. This one-at-a-time action is illustrated in Figure 11-1.

Asynchronous System

The *asynchronous* I/O system requires a separate control unit with many of the features of a central processor. It has *storage* facilities, can make limited *decisions*, and can contend with various *speeds* and *formats*. A sample layout of such a system is illustrated in Figure 11-2.

Here the I/O controller appears to be the center of the system. It is the *focal* point of all I/O operations. The various peripheral devices are under the *direct control* of the I/O controller. The controller performs *no* action on its own initiative; it does only what it is *told* to do.

Our central computer has an arithmetic module and two memory modules. Of course, our usual control logic is contained in each of these modules. The I/O operation is *initiated* by an instruction which says, "Send C(memory location XXX) to I/O controller." The contents of the specified memory location is an I/O operation *control word*. It *identifies* the type of operation (input or output), the peripheral device to be used, the number of words to be transferred, and specifies the starting memory address for reading or writing the data. Once the control word has been sent to the controller, the central computer is *free* to continue its computing task.

The controller follows the exact instructions in the control word. It *activates* the designated peripheral device, *locates* the desired information, *transfers* the data in a manner suitable to both the central memory and the peripheral device, *supervises* the entire operation, and *signals* the central computer when the task is finished.

An I/O controller module may contain *several* individual I/O controller units. *Each* unit is capable of performing a *separate*, and completely independent, I/O operation. One may be performing an input operation from magnetic tapes while another is outputing via the printer or card punch. In fact, several controllers may utilize the services of a *single* peripheral device on a time sharing basis.

In the *asynchronous* system, the central computer *initiates* the I/O operation when directed by the program. It then performs *other* functions. The I/O controller *notifies* the central processor when the new data is ready or when the output operation is finished. The controller *requests access* to the designated memory module when each computer word is ready for transfer. A *memory cycle* is still essential for each word being moved, but these extra cycles have no significant bearing on the normal computing actions.

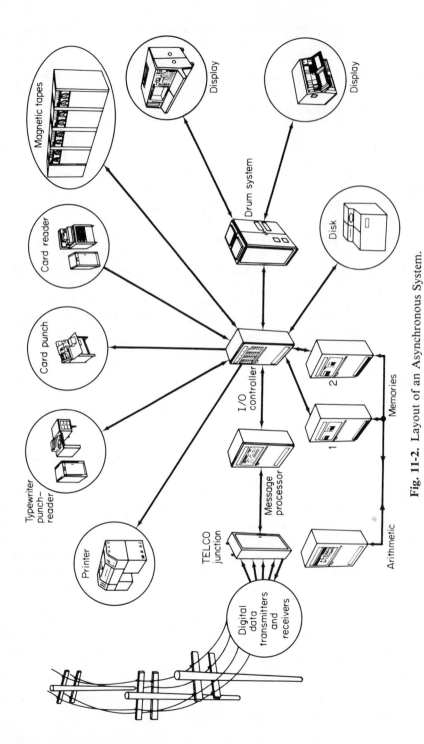

Fig. 11-2. Layout of an Asynchronous System.

Magnetic tapes

Display

Display

Card reader

Card punch

Disk

Drum system

Typewriter
punch-
reader

I/O
controller

Memories

2

1

Message
processor

Printer

Arithmetic

TELCO
junction

Digital
data
transmitters
and
receivers

The system can be *expanded* to include as many pieces of individual equipment as the user's need dictates. It is conceivable that a single system may contain 2 arithmetic sections, 20 memory cabinets, 6 I/O controllers, 2 drum systems, 10 display consoles, 6 disk units, and 2 message processors with *each* controller operating 20 or more peripheral devices.

The *overlapping* and time saving functions of the completely asynchronous system are illustrated in Figure 11-3.

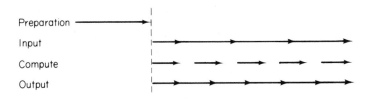

Fig. 11-3. Actions in an Asynchronous System.

In this system, some time is lost on *preparation*, and an occasional *break* in computation occurs when the central computer must wait for information. However, as illustrated by the graph, more of the computer operating time is *productive*. You may well ask why all systems are not asynchronous.

It is simply a matter of economics. A firm purchases a system to fit its need. If it can afford less efficiency in operating time, the system will contain fewer pieces and less elaborate hardware. The larger the organization, the more valuable the productive time, and the more elaborate the hardware.

Some systems are completely *synchronous;* one operation at a time. Others are completely *asynchronous;* several operations time shared to appear simultaneous. Most systems, however, have been designed around a *compromise* between the highest possible efficiency and lowest feasible cost. This means that many systems will have *some* I/O operations that are synchronous and *others* that are asynchronous.

Some of the more *elaborate* systems must have nearly 100% computing time. These systems make use of dual hardware and extensive executive programs. One system is *on line* (under the control of the main executive program) while the other is *off line* preparing data, runing maintenance diagnostic checks, and exercising the redundant equipment. These systems are most frequently encountered in such operations as space probes, national defense, and other places where cost is incidental.

11-2 CATEGORIES OF PERIPHERAL DEVICES

One good reason for combining inputs and outputs into the same chapter is the construction of the peripheral devices. Some of them are for *inputs only;*

some are for *outputs only;* and some perform *both* input and output operations. We group the devices into *two* categories; *simple* and *complex.*

Simple Devices

A simple peripheral device is capable of performing only *one* type of operation. This can be *either* input or output. Input is interchangeable with *read*, and output is interchangeable with *write*. For instance, during a *read* operation, information is passing *from* the input device *to* central memory. During a *write* operation, information is passing *from* central memory *to* the output device.

Some *examples* of simple devices are card readers, card punches, and line printers. There are *two types* of card punches. One is for off line use and is manually operated from a keyboard. This is *neither* an input nor an output device. It is simply a machine to aid in the preparation phase. The second type of punch is *on line* device, and it is under program control. This automatic punch is a simple *output* device which stores information from memory by punching it into cards.

The card reader is a simple *input* device which moves information from the punched card to central memory storage. The operator controlled punch can use an *indefinite* period of time preparing and storing information on punched cards. The cards can then be placed into the card reader, which is under program control. At the appropriate spot in the program, the cards will be read, and the information will be stored in memory. Extensive batches of data and many complete programs are loaded into central memory through the card reader.

The line printer is a simple *output* device. It is operated on line and under program control. Many printers have the capacity for an 80 to 132 character line. They are relatively high speed since they print a complete line in a single action. The printer output may be used as a permanent hard copy record as well as a copy for checking data.

Complex Devices

Complex devices are used for *both* input and output operations. Some *examples* of complex devices are tapes, drums, keyboards, and disks. The keyboard provides a direct link between the operator and the central computer. The operator can type messages which go directly into memory. He may also interrupt the computation process and have his inputs processed immediately. The computer's replies are typed out on the same keyboard under program control.

Combination keyboard and CRT display units are popular for remote control stations. In a *time sharing* system, several of these remote stations are tied into a single central computer. Examples of their use are *banking* and computer assisted *education* facilities. It is conceivable that thousands of students can use the same computer, study many different courses, and be separated by thousands of miles. Each user has the impression that he has the computer's undivided attention.

Magnetic disks, drums, and tapes are generally on line devices under program control. This means that they can be used to obtain data when it is needed by the central processor. They are also on *standby* to store any information that needs to be moved out of the central memory.

A drum system generally provides an *intermediate* link between the central computer and the display system. Display information is stored on the *drum* and updated as necessary by program control. This information is then transferred from the drum to the *display* device in a repetitive fashion.

The magnetic drum is sometimes used as a buffer for *temporary storage* of messages going out and coming in over transmission lines. If the drum is not used, the buffering is provided by a unit which we call the message processor. These *buffering* devices receive data from the central processor in *parallel* form (a full word at a time) and send data over the line in *serial* form (one bit at a time). Input messages arrive over the lines in serial form and go to the central memory in parallel form. A buffering device has the capability of processing *many* messages, on a time sharing basis, without either delaying the messages or losing bits. The remote stations previously described have *access* to the central computer through telephone lines and a buffering device. Each *remote* station probably uses a separate line, but all messages may be processed by the *same* buffering device.

11-3 COMMON PERIPHERAL DEVICES

Without peripheral equipment, the central computer is a piece of expensive, but *useless*, hardware. The peripheral equipment provides for a two way flow of information and completes the system. In this section we will take a brief look at the *applications* of some of the most common peripheral devices.

Control Unit

Most types of peripheral devices have been in use for a long time. It is not necessary for the designer of a new system to *create* each piece of peripheral gear. He can *select* as many of these as he needs from the open market. Several companies specialize in manufacturing only peripheral devices. These off-the-shelf items can be made compatible with most central processors.

Each type of peripheral device is accompanied by a *control* unit. One control unit, or controller, may handle *several* devices of one type. These controllers bridge the communications gap between the peripheral device and the central computer. They contain the compatibility circuits which make it possible for a TELEX* tape drive to function in harmony with an IBM central computer. *(TELEX is a registered trademark of TELEX Computer Products, Inc.)

Card Reader

The card reader is a very simple machine. It moves punched cards from a hopper to a stacker, and reads the information that has been recorded on the card. Some readers move the cards from roller to roller along a level path as shown in Figure 11-4. Others move the cards along a semicircular path similar to the illustration in Figure 11-5.

Both of these illustrations show a *contact* roll and a *wire brush* as the reading mechanism. (There are other ways to read the card, photoelectric cells for example). In the brush-contact method there are 80 brushes, one for each column. All 80 columns are read simultaneously. A *hole* in the card allows the

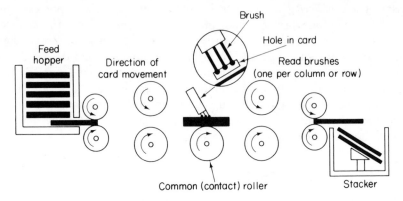

Fig. 11-4. Cards Moving on a Level Path.

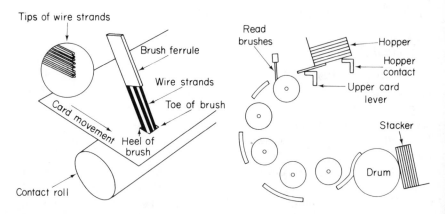

Fig. 11-5. Cards Moving in a Semi-Circle.

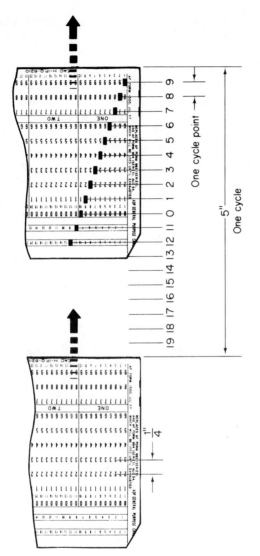

Fig. 11-6. Card Movement.

metal brush to drop through and complete the electric circuit with the metal contact roller. *No hole* means no contact.

The card *moves* through the machine face up, edgewise, and row 9 first. This is illustrated in Figure 11-6.

The distance between the leading edges of two consecutive cards is 5 inches. This distance is broken into 20 distinct timing periods which we call *cycle points*. The total distance is one *card cycle*. The translation circuits interpret the information according to the cycle points where contact was made.

At 150 cards per minute, the *time* between cycle points is 2 ms. The information from all 80 columns and 12 rows goes into a storage register when the 12th row is read. Central memory has 8 cycle points (16ms) to move this data into core storage before the next card is in the read position. Using a $6 \mu s$ memory cycle and moving one word per cycle, this gives us time to move more than 2600 words.

As the cards approach the stacker they are attached to a *stacking drum* which deposits them in the stacker. This action is illustrated in Figure 11-7.

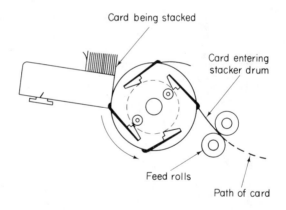

Fig. 11-7. Stacking Cards.

The drum speed is synchronized with the card movement so that a spring loaded *gripper* is always in position to intercept the edge of the card. The feed rolls force the edge of the card under the gripper on the stacker drum. As the gripper passes the stacker, the card is *stripped* away and deposited in the stacker.

Card Punch

The keyboard operated punch has the capability of *reading* a master card and punching the identical information into any specified *number* of cards.

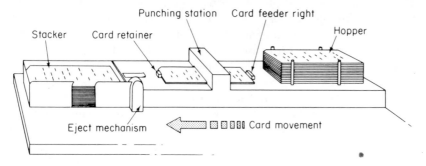

Fig. 11-8. Cards Moving Through Punch Station.

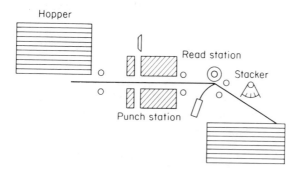

Fig. 11-9. Side View of Punch Station.

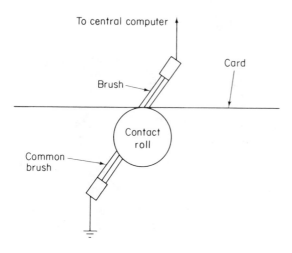

Fig. 11-10. Read Station in Card Punch.

The computer operated punch can *read* a card and punch the identical information into the *following* card. Both punches, of course, have the ability to *ignore* the reading station and record different information on each card. Figure 11-8 illustrates card movement from hopper to stacker.

The cards are stacked in the hopper face up with column 1 in front. As they move through the punch station, all forward motion ceases as each column is under the punch. There is a punch for each of the 12 rows. The *activated* punches are driven through the card and withdrawn while the card is stationary. The side view in Figure 11-9 shows a punch station with one punch poised in position.

The read station is exactly *one* full cycle after the punch station to facilitate reading one card and punching the next. The reading action is similar to that described for the card reader. A better view of the read station is shown in Figure 11-10.

Paper Tape Recorder

Like the card punch, the paper tape recorder comes in two types: *keyboard* operated and *computer* operated. Many keyboards are designed for manual operation by the operator and automatic operation by the computer. Most have a paper tape reader and recorder built in, but there are separate

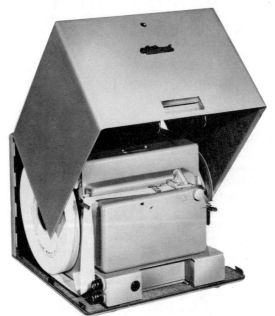

Fig. 11-11. Paper Tape Recorder. (Courtesy National Cash Register)

devices designed for recording information on paper tape. One such is shown in Figure 11-11.

This computer operated machine is smaller than a four inch cube when the cover is closed. The blank tape is shown on a roll. The perforated tape, coming from the right side, is wound onto a reel. Figure 11-12 shows a section of perforated tape.

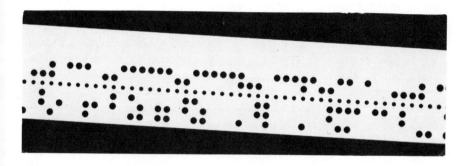

Fig. 11-12. Section of Perforated Tape. (Courtesy National Cash Register)

Paper Tape Reader

There are several ways of *sensing* the holes in paper tape during the reading process. Three of these are illustrated in Figure 11-13.

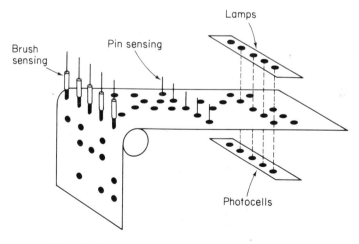

Fig. 11-13. Methods of Reading Perforated Paper Tape.

Paper tape readers vary from very simple to rather elaborate machines. A high speed photoelectric reader is shown in Figure 11-14.

This is one of the more sophisticated readers. It can be operated on line or off line. Data can be read *automatically* into the system, and areas of data can be *addressed*. It has an automatic *rewind* feature which keeps all the information available at all times.

Magnetic Tape

A magnetic tape drive is both a reader and recorder. It can be operated on line or off line, and data on a reel is addressable. The computer can *write information* on the tape as well as *read* information from the tape. Figure 11-15a is a tape *drive* and b is the *controller*.

Modern tape controllers, such as this, can handle up to 8 tape drives simultaneously. The tape drives feature automatic threading, speeds in excess of 200 inches per second, and data density of more than 1600 bits per inch.

A more advanced approach is to package both tape drive and tape controller in the same cabinet as shown in Figure 11-16.

This combination features two tape drives and a controller in one package.

Fig. 11-14. Paper Tape Reader. (Courtesy National Cash Register)

Fig. 11-15. Tape Drive and Controller. (Courtesy TELEX Computer Products)

Fig. 11-16. Tape Subassembly. (Courtesy TELEX Computer Products)

The controller can still handle up to *six* tape drives in any combination of compatible models.

Line Printer

The line printer is the *fastest* method yet devised for moving data from the central memory to printed copy. These printers print an entire line of type at one time. The paper moves in a rolling motion and pauses imperceptibly for each line to be printed. Several methods have been used to *print* the characters: print wheels, wire matrices, and print chains are a few.

Most printers feature 48 different characters and 120 characters per line of print. Continuous print speeds in excess of 1200 lines per minute have been attained.

Users are not limited to the 48-character sets of printing symbols. There is a 288-character type array to choose from, and lines may be expanded to 132 characters. Generally, the larger the type set the slower the speed. One printer by TELEX features 1200 lines per minute with a 48-character set, 1500 lines per minute with a 36-character set, and burst modes of 2500 lines per minute with a 16 character set.

Some printers can *accept* inputs from magnetic tape units as well as from

Fig. 11-17. Off-Line Train Printer System. (Courtesy TELEX Computer Products)

the central computer. This enables either on line or off line operation. Data from core memory can be written onto magnetic tapes at speeds far in *excess* of the printer's capabilities. The tape-printer combination *saves* valuable computer time. The computer writes the output data onto tape; the tape unit then *operates* the printer. The *difference* between maximum tape speed and maximum printer speed can be converted directly into computer *time saved.* Figure 11-17 is a tape-printer combination.

Additional tape units and printers can be added without increasing the number of control units.

Magnetic Disks

Magnetic disks provide high speed *access* to almost unlimited data storage locations. Information can be *written* onto and *read* from disks by the central computer. New data can be written *over* the old and data remains until the area is rewritten. One disk controller can *operate* up to 9 disk storage drive units. Figure 11-18 shows a controller with a group of storage units.

This picture features the TELEX 5600 *double density* disk storage system. It has *fast* start and *rapid* access capabilities. Disk packs can be removed and replaced at will, providing unlimited capabilities for a data library.

Fig. 11-18. Disk Storage Drive Units, with Controller. (Courtesy TELEX Computer Products)

Display Devices

Display devices cover a range from a simple neon glow tube to elaborate consoles and projection systems. There are display tubes to fit the needs of nearly any conceivable use. The military uses *elaborate* displays to picture all aircraft within a given area. In computer assisted instruction, the student uses

a comparatively *simple* graphic and alphanumeric display. Figure 11-19 shows
a display unit of the latter type.

With a unit such as that in Figure 11-19, there is *two way* communication

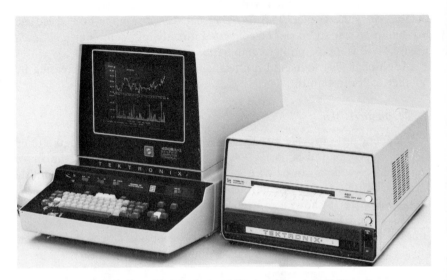

Fig. 11-19. Graphic Display and Copier. (Courtesy Tektronix,
Inc)

between the computer and operator. The *operator* types questions, answers,
or information requests on the keyboard. The *computer* answers by displaying
printed information or pictures. The operator can obtain hard copies of any
displayed information when he so desires. The keyboard is similar to that of
a standard typewriter. The computer can be programmed to accept *plain
English* inputs and reply in kind.

This particular model is very popular with users of a *time shared* computer
system. A *student* can receive personalized instruction from a programmed
text, or a bank *teller* can call up a specific page of a ledger.

Review Exercises

1. Why are peripheral devices essential to a computer?
2. What type of devices perform:
   a. Encoding?
   b. Decoding?
3. Describe the relation between total input/output time and total system operating time.

4. Name and describe the two basic systems for timing and control of I/O operations.

5. How does the I/O controller determine the details of a directed I/O operation?

6. A system has three I/O controllers and 20 peripheral devices. (a) What is the maximum number of I/O operations that can be performed simultaneously? (b) Why?

7. Name the two categories of peripheral devices and describe the meaning of each.

8. Indicate the category of each of the following devices:
a. Line printer.
b. Magnetic drum.
c. Card punch.
d. Magnetic tape.
e. Card reader.
f. Keyboard.

9. Identify the devices in item 8 according to their function as input, output, or both.

10. Which peripheral device is most often used as a buffer between the central computer and the display unit?

11. What is the meaning of:
a. Card cycle?
b. Cycle point?

12. Describe the position of a card as it moves through the:
a. Reader.
b. Punch.

13. List three methods of sensing the holes in perforated paper tape.

14. What is the chief advantage of a magnetic tape drive over a perforated paper tape drive?

15. What is the usual maximum load for a magnetic tape controller?

16. What is the fastest method of converting central memory data into hard copy?

17. What is the chief advantage of combining magnetic tapes with line printers? Explain.

chapter twelve

instruction
analysis

Regardless of your reason for studying computers (systems maintenance, programmer, analyst, etc.), the *most valuable* knowledge that you can obtain is the timing and control of a system as it *executes* individual instructions. The purpose of this chapter is to develop that knowledge.

Something more than a hypothetical system is needed for this vital phase of computer principles. The system should be fairly complex in order to promote a *transfer of knowledge* to other systems that you may encounter. The system we will use for this purpose is the *AN/FSQ-7*, Semi-automatic Ground Environment (SAGE) system which is used by the U.S. *Air Force*. This system was built by the *IBM Corporation*.

The block diagrams, instructions, timing charts, and functional commands in this chapter have been created by *modifying* those for the AN/ FSQ-7. Modifications and reproduction of this information were made possible by special permission of the IBM Corporation.

319

12-1 SYSTEM ORGANIZATION

The system has been simplified considerably but not enough to prevent its functioning as a system.

Block Diagram

Figure 12-1 is a block diagram of the central computer of our *modified* system. Notice that it is only slightly more complex than the sample system we used previously.

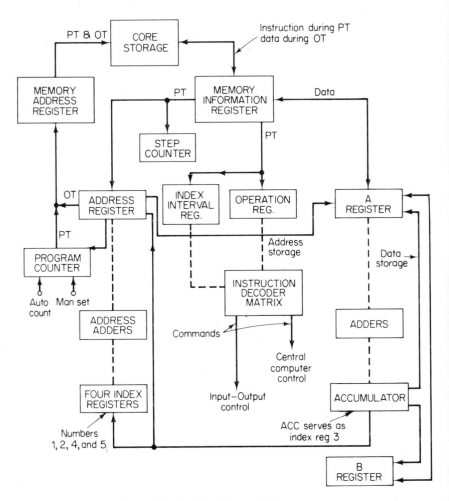

Fig. 12-1. Modified Central Computer.

The *new* blocks on this diagram are the step counter, the index interval register, and the B register. We have already encountered the actions of both the step counter and the B register in our multiply and divide instructions. The index interval register is actually an *extension* of the operation register. Some instructions have several variations. The bits which specify the variation are hela by the index interval register for decoding.

This is a stored program, single address machine, and all input/output operations are synchronous in nature.

Timing

Timing is the same as previously discussed. There are *three*, 6 μs machine cycles: PT, OTA, and OTB. Each of these is divided into 12 pulses by a ring counter. Consecutive timing pulses (TPs) are separated by 0.5 μs at the leading edges. Each *machine* cycle is supported by a 6 μs *memory* cycle. The functions of each cycle (memory and machine) are the same as those previously discussed.

Common Commands

Commands for timing and control are generated by an instruction decoder matrix in a manner similar to that described earlier.

Several commands occur as *standard* procedure at specific times during a given machine cycle. For instance: at TP7 of each program time cycle, command 92 *adds one* to the contents of the program counter (previously called program address counter). This is the action which creates the *address* of the next instruction in the program sequence. Since this action is *common* to all PT cycles, we call command 92 a *common command*. All the common commands are shown on the chart in Figure 12-2.

Across the *top* of this chart, we have the timing pulses for all three machine cycles; PT, OTA, and OTB. The PT-OT flip flop determines if our cycle is to be PT or OT. If this flip flop is *clear*, our next cycle will be *PT*; if it is *set*, our next cycle will be *OT*. When OT is selected, the A-B flip flop determines whether we have OTA or OTB. If the A-B flip flop is *clear*, out cycle is *OTA*; when it is *set*, we have OTB.

Notice the *sections* of the chart labeled A, B, and C. Section A includes only *program time*. Section B is program time *plus* operate time A. Section C includes *all three* cycles: PT, OTA, and OTB. The instructions are divided into categories to match these three sections. The *A category* instructions use all the common commands indicated in section A of the chart. The *B category* instructions use the common commands in section B. The *C category* instructions use all the common commands in section A and C, and sometimes, all three sections.

322

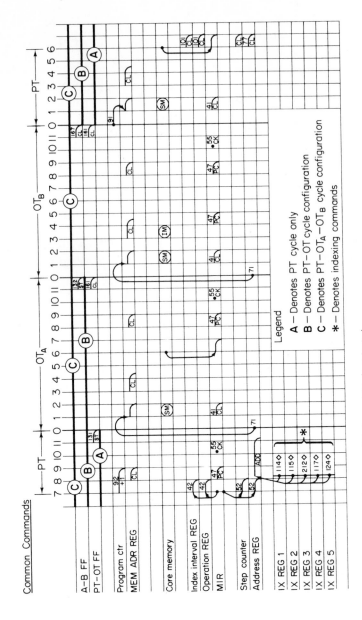

Fig. 12-2. Common Command Timing Chart.

We can extract a great deal of information from this chart with a minimum of effort, and reading it is a skill easily acquired. Scattered about on the chart, we have several numbers. These are the numbers which identify *specific commands.* In the vicinity of each command number, we have a line, an abbreviation, or a symbol which tells us what that command does. The *position* on the chart tells us where the action takes place and at what time. A few actions take place by routine timing action without generating a command. Two of these appear on the chart: pulses which *clear* the memory address register at TP3 and 8 of each cycle, and the transfer signal which *moves* information from core memory to the memory information register at PT6 and OTA6.

In addition, we have two commands that are identified by *abbreviations* in a circle. The encircled SM *starts* a memory cycle at TP1 of each machine cycle. The encircled IM occurs only during the OTB cycle. IM means *inhibit memory.* It prevents transfer of C(core register) to MIR during the read portion of the OTB cycle. This is the action which *clears* the core location to receive a new word.

Common Commands List

41—Clear memory information register
42—C(MIR) to operation and index interval registers
47—Parity count
52—C(MIR) to address register and step counter
55—Parity check
71—C(address register) to MAR
77—Clear address register
91—C(Prog. Ctr) to MAR
92—Add one to program counter
101—Clear operation register
114—C(Ix Reg 1) to address register
115—C(Ix Reg 2) to address register
117—C(Ix Reg 4) to address register
124—C(Ix Reg 5) to address register
131—Set PT-OT flip flop to PT
132—Set A-B flip flop to B
161—Clear PT-OT flip flop to PT
167—Clear A-B flip flop to A
212—C(Ix Reg 3) to address register
IM—Inhibit memory output
SM—Start memory cycle

The lines on the chart are easy to read, and they are very informative. The line *begins* in the area where the action starts, and at the *time* the action starts. The line then *traces* the action to the point of *destination* and terminates when the action is completed. Here are some examples:

1. At PT0 in the program counter, we have command 91. Command 91 *moves* C(Prog. Ctr) into the memory address register. The transfer is completed by PT2.

2. At OTA0 and OTB0, we have command 71 in the address register. It *moves* C(Add Reg) into the MAR and has the transfer completed by TP2 of the next cycle.

3. At PT7 we have two commands in the memory information register. Command 42 *transfers* C(MIR) into the operation register and the index interval register. At the same time, command 52 *moves* C(MIR) into the step counter and into the address register.

4. At TP8 of each cycle, we have a copy of the transferred memory word still in the MIR. Command 47 performs a *parity count* on this word.

The common commands list provides a concise *description* of the action accomplished by each common command. Study it with the chart; the same information is in both places.

12-2 INSTRUCTIONS

From a list of 61 *basic* instructions and more than 200 specific *variations*, we have composed a list of 31 basic instructions with a very limited number of variations.

Mnemonic Code

The *mnemonic* code is a three letter instruction identification that is suggestive of the function and easy to remember. Some examples are ADD for add, MUL for multiply, STA for store, and SLR for shift left end round.

Octal Code

Since the computer system deals only in numbers, we need a *numeric* code for each instruction. Some of the instructions on our list have a *two* digit octal code. When that is decoded in the operation register, the configuration of the six binary bits will identify the instruction and create the commands necessary for execution.

Classes

The instruction list is divided into eight classes. As might be expected, all instructions of a class have several points in common. For instance, the only difference between add and subtract is *one* command. The subtract instruction generates command 26 to *complement* the C(A Reg) just before it is added to C(Acc.).

Variations

Several of the instructions have a *hyphen* (-) before the octal code. This indicates that the code is incomplete. One of the *index interval codes* is to be inserted in place of the hyphen.

Following is a list of instructions. The instructions are identified by both mnemonic code and octal code, and they are divided according to class. A list of index interval codes follows the instruction list.

Instructions List

MISC. CLASS	MULT. CLASS	BRANCH CLASS	STORE CLASS
HLT 00	*MUL —25	BPX —51	*STA —34
PER —01	*DVD —26	BFZ 52	*ECH —35
SLR 02		BRM 54	
*LBD —03	*I-O CLASS*	*ADD CLASS*	*SHIFT CLASS*
*CMF —04	*LDC —60	*CAD —10	DSL 40
TOB —05	*SDR —61	*ADD —11#	DSR 41
TTB —06	SEL —62	ADB 12#	ASL 42
	*RDS —66	*CSU —13	ASR 43
	*WRT —67	*SUB —14#	DCL 46

RESET CLASS
X1A —75
XAC —76

   *   Indexable instructions.

   #   Instructions which can cause overflow.

   —   Instructions which may use index interval bits.

The functions of the instructions are:

HLT—Stops the computer.

PER—Operates a selected device.

SLR—Shift C(Acc & B Reg) left a specified number of spaces then round off the number to the most significant 16 bits.

Instructions List

LDB—Load the B register with a word from the specified memory location.

CMF—Compare the full word from specified memory location with the full word in the accumulator.

TOB—Test a specified bit. If the bit is a 1, skip the next instruction.

TTB—Test two bits. If 1st bit is a 1, skip one instruction. If 2nd bit is a 1, skip two instructions. If both bits are 1s, skip three instructions.

CAD—Clear the accumulator and move the word from the specified memory location into the accumulator.

ADD—Add C(specified Mem. Loc.) to C(Acc).

ADB—Add C(B Reg) to C(Acc).

CSU—Clear accumulator and add to it the complement of the word from specified memory location.

SUB—Subtract C(specified Mem. Loc.) from C(Acc).

MUL—Multiply C(specified Mem. Loc.) times C(Acc).

DVD—Divide C(Acc & B Reg) by C(specified Mem. Loc.).

STA—Store C(Acc) into specified memory location.

ECH—Exchange C(Acc) with C(specified Mem. Loc.).

DSL—Shift C(Acc & B Reg) a specified number of positions to the left.

DSR—Shift right C(Acc & B Reg) a specified number of positions.

ASL—Shift left C(Acc) a specified number of positions.

ASR—Shift right C(Acc) a specified number of positions.

DCL—Double cycle C(Acc & B reg) a specified number of positions to the left.

BPX—Branch to specified location when sign of specified index register is positive. If no register is specified, or specified register is negative, go on to the next instruction.

BFZ—If C(Acc) is zero, branch to specified location. Otherwise go on to next instruction.

BRM—If sign of accumulator is negative, branch to specified location. Otherwise go on to next instruction.

LDC—Load address counter with specified number.

SDR—Select drums.

SEL—Select specified device.

RDS—Read a specified number of words.

WRT—Write a specified number of words.

XIA—Load C(Add. Reg) into the specified index register.

XAC—Load C(Acc) into the specified index register.

Index Interval Codes

Operate		Select	
Condition lights	1–4	Camera	1
Inhibit alarm	5	Drum	2
Clear I/O interlock	6	Printer	3
Start display	7	Display	4
Start printer	10	Tapes	5
Set tapes prepared	11	Card reader	6
Rewind tapes	12	Card punch	7
Read tapes	13	Keyboard	10
Write tapes	14		
Start card reader	15		
Start card punch	16		
Enable keyboard	17		

Word Format

We are dealing with a 16 bit word. The instruction word *is not* a signed quantity, and this releases the sign bit for other purposes. The division of our instruction word is shown in Figure 12-3.

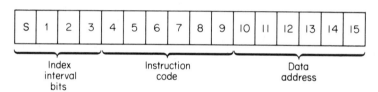

Fig. 12-3. Instruction Word.

Notice on the list of instructions that certain instructions can be *indexed.* This means that the number in a specified index register may be used to *compute* the data address. With these instructions, bits 1, 2, & 3 contain the number of the index register to be used. If these bits specify Ix Reg 1, 2, 3, 4, or 5, indexing *will* occur, and the data address (bits 10-15) are added to C(Ix Reg). When bits 1, 2, & 3 contain 0, 6, or 7, indexing *is not* indicated. In this case, the address specified by bits 10-15 is used.

We have five 7-bit index registers to choose from, and index register number 3 is composed of the sign and high order six bits of the *accumulator.*

When we use the contents of an index register, we have an *option* of incrementing the number by one or decrementing it by one. We express our option with the *sign bit* of the instruction. A *zero* sign bit increments; a *one* sign bit decrements. For instance: when bits S, 1, 2, and 3 are 0101, we add C(Ix Reg 5) to the specified address and add one to C(Ix Reg 5). Suppose that Ix Reg 5 contains 0.011 111 (+37) and our instruction word is 0.101 001 001 001 000. The instruction says, "add the data from location 10 + 37 to C(Acc)." After the address is computed, we access memory location 47, and the number 0.100 000 (40) is returned to Ix Reg 5. (37 + 1 = 40).

Notice on the instruction list that we have *seven hyphenated* instructions that are *not* indexable: PER, SEL, BPX, TOB, TTB, XIA, and XAC. PER is operate; SEL is select; and BPX is branch. The *operate* section of the index interval codes indicates the operation for various configurations of bits S, 1, 2, & 3 when using the PER instruction. The *select* portion of the index interval codes is used with SEL.

The BPX uses the index interval code bits to specify the *conditions* of the branching action. If the interval code specifies 0, 3, 6, or 7, the program will branch *unconditionally*. If these bits specify index register 1, 2, 4, or 5, branching will occur only if the selected register contains a *positive* number.

TOB is test one bit; TTB is test two bits. With these instructions, the index interval code specifies which *bits* are to be *tested*. The bit is specified by the code 0000 to 1111 which corresponds to word bit positions S and 1-15. The TTB checks the *specified* bit and the *next* lower order bit. If TTB 1111 is specified, bits 15 and S are checked.

Several instructions have *no need* for a data address. In this case, bits 10-15 are used for other purposes, or they are simply ignored by the machine. A case in point is any shift instruction. The address portion of a shift instruction specifies the number of shifts.

Unique Commands

On the following pages, we have *timing charts* for each of these instructions along with explanations. We will find that a particular instruction uses certain *common* commands and some *unique* commands. Following is a list of the *unique* commands. They are listed in *numerical* order for ease of reference. For the moment, just glance through the list to become familiar with the nature of the contents. You will need to *refer* to it frequently as we discuss timing and control.

Unique Commands List

COMMAND NUMBER	ACTION ACCOMPLISHED
1—	C(Acc) to B register
2—	Shift Acc 2-15 to 1-14
4—	Shift Acc 1 to sign

Unique Commands List

5—	Move Acc sign to B register 15
6—	Move Acc sign to Acc 15
7—	Move Acc 15 to B register sign
9—	Correct sign
10—	Clear Acc
11—	Correct remainder
13—	Make Acc positive
16—	Make Acc and B register positive
17—	C(Acc) to MIR
18—	Shift Acc right
19—	Complement C(Acc)
21—	Clear A register
22—	Make A register positive
23—	C(A register) to MIR
26—	Complement C(A register)
39—	C(MIR) to B register
43—	C(MIR) to A register
60—	Shift Acc left
61—	Make signs of Acc and A Reg unlike.
62—	Carry 1 to LSB adder
63—	End carry 1 to LSB adder
64—	Carry zero to LSB adder
66—	Record overflow
72—	C(Add Reg) to I/O address register
73—	Subtract one from C(step counter)
74—	Set step counter to 17_{10}
75—	Set step counter to 15_{10}
77—	Clear address register
78—	C(Add Reg) to Ix Reg 2
79—	C(Add Reg) to Ix Reg 1
80—	Compute partial quotient
81—	B register sign to Acc 15
82—	Shift B Reg bits 1-15 to S-14
83—	Compute partial product
84—	Clear B register
85—	Shift B Reg S-14 to 1-15
87—	Carry 1 to LSB adder
88—	C(B Reg) to A register
89—	C(B Reg Store FF) to B Reg sign
90—	Add two to C(Prog. counter)
92—	Add 1 to C(Prog. counter)

Unique Commands List

93—	C(Prog. counter) to A register
94—	Clear program counter
95—	C(Add Reg) to Ix Reg 4
96—	C(Add Reg) to Ix Reg 5
99—	Compare C(Acc) and C(A Reg)
100—	C(Add Reg) to MAR
102—	Select specified bit or bits for testing
104—	Sense condition of operate gates
113—	Clear Ix Reg 1
114—	C(Ix Reg 1) to address register
115—	C(Ix Reg 2) to address register
116—	Clear Ix Reg 2
117—	C(Ix Reg 4) to address register
118—	Clear Ix Reg 4
119—	Test C(Ix Reg) for positive
124—	C(Ix Reg 5) to address register
125—	Clear Ix Reg 5
126—	Test C(Ix Reg 5) for positive
134—	Set pause flip flop
138—	Set 2 MHz sync flip flop
144—	Clear I/O register
146—	Clear drum control register
148—	Clear I/O address counter
149—	Clear I/O word counter
151—	C(Add Reg) to drum control register
152—	C(Add Reg) to I/O word counter
154—	C(Add Reg) to program counter
155—	Deselect pulse
156—	Select pulse
162—	Test accumulator sign for a negative
163—	Clear branch flip flop
164—	Test C(Ix Reg 1) for positive
165—	Test C(Acc) for negative
170—	Set branch flip flop
174—	Test C(Ix Reg 2) for positive
180—	Start read
182—	Start write
214—	Accumulator sign correction
270—	Clear continue flip flop
325—	Select pulse for drums

12-3 TIMING AND CONTROL

Since we are interested primarily in stored program computers, an understanding of the *timing and control* during execution of specific instructions is the *key* to understanding the machine. The timing charts of the individual instructions contain all the information. A *clear* understanding can be developed by studying the charts in conjunction with the list of unique commands and the system block diagram. We'll take the instructions in the order of their octal codes.

Halt

The *half* instruction is a category B instruction. It requires a PT cycle, an OTA cycle, and sometimes an I/O pause. The mnemonic code is HLT, and the octal code is 00. This instruction *stops* the computer. The details of how this is done are shown on the timing chart in Figure 12-4.

We will discuss the halt chart in some detail since it is our first. All instruction executions *begin* at PT7; PT0 through 6 are always used to bring the instruction from core storage to the memory information register.

Glance across the *top* of the chart. The *common* commands are listed, at the proper times, above the first horizontal line. Between PT7 and 8 we have common commands 42, 52, 47, and 92. *42* moves the bits S-4 into the index interval register and bits 5-15 into the operation register. *Command 52* moves bits 10-15 into the address register and the same bits into the step counter. *Command 47* performs a parity count on the instruction word. *Command 92* adds one to the C(Prog. Ctr).

Between PT10 and 11, command 55 *checks* the result of the parity count. If it is *odd*, action continues; if it is *even*, the instruction is invalid and action ceases. At PT11, command 131 sets the PT-OT flip flop to make the next cycle OT.

The I/O *pause* happens only if we try to stop during an I/O operation. There are no common commands during the pause. If the pause occurs, it lasts only until the current I/O operation is completed. Back on the common command chart (Figure 12-2), we have done *nothing* to the A-B flip flop. This means that it is *clear*, and our next timing pulse is OTA0. This is true either with or without the pause.

Between OTA1 & 2, we have commands 41, 71, and SM encircled. 41 clears MIR; 71 moves the contents of the address register to MAR; and SM starts another memory cycle. Between OTA7 & 8, command 47 performs a parity count on the word in MIR. This parity count is checked at OTA10 by command 55.

At OTA10, command 161 *clears* the PT-OT flip flop to make the next cycle *program time*. Even though we are terminating automatic action, the

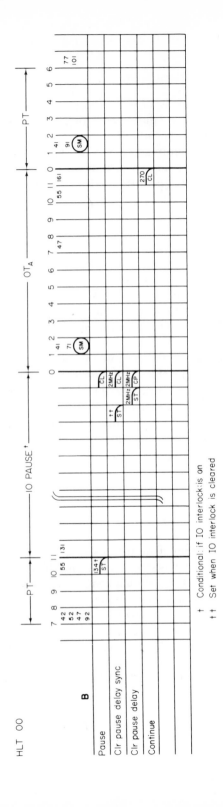

Fig. 12-4. Halt Timing Chart.

next instruction will be brought from core storage to MIR before the machine stops. When the machine is restarted, (provided power is not removed) the instruction is *ready* to decode.

With the halt instruction, no useful work is accomplished during OTA by commands 41, 71, SM, 47, or 55. We started the cycle *in case* we need an I/O pause. Once started, it must run its course of 6 μs (with or without pause). The commands are common to this cycle, and are generated as a matter of course. As OTA terminates, *command 161* brings us back to program time.

At PT1, we have commands 41, 91, and SM. *91* moves the next instruction *address* from the program counter to MAR; *41* clears MIR; and *SM* starts another memory cycle.

Back on the common command chart at PT6, the contents of the selected core *location* are automatically transferred to MIR. Commands 77 and 101 also occur at PT6. *77* clears the step counter and the address register. *101* clears the index interval register and the operation register.

All halt instruction actions, (except the conditional I/O pause) described thus far, are common to *all* category B instructions. Each of these actions is clearly depicted on the common command *chart*. This gives us the general picture for category B instructions. When we add a few *unique* commands, we have the *halt* instruction.

In the body of the HLT chart, we have *command 134* at PT10. The chart tells us that this command occurs only if the I/O *interlock* is on. If this interlock is on, it means that an I/O operation is *in progress* and a *pause* is required. The pause is initiated by 134 setting the pause flip flop. If the I/O interlock is *clear, no pause* is needed, command 134 *is not* generated, and timing moves us directly to the OTA cycle.

If we do pause, the *duration* is unpredictable; it will last until the I/O operation reaches a natural *conclusion*. When the I/O interlock *clears*, a signal sets the clear pause delay Sync. A half μs later, a 2 MHz signal sets the clear pause delay. When the clear pause delay is *set*, the pause will terminate 1/2 μs later. During the last *half* μs, we have a 2 MHz pulse which *complements* the clear pause delay from set to *clear;* the same 2 MHz pulse *clears* the clear pause delay Sync; and another signal (unidentified) *clears* the pause flip flop. The ring counter *resumes* action, and the next clock pulse generates OTA0 just as if nothing had happened.

No unique commands occur during OTA until we reach TP11. At this time *command 170* clears the *continue* flip flop. Timing will cease exactly 3 μs after the continue flip flop is cleared. This gives the machine time to go through the first *half* of a PT cycle, which we need to bring the next *instruction* from core storage to the memory information register.

Operate

The *operate* instruction generates a control pulse to operate the equipment which we specify by the index interval bits. Figure 12-5 is the timing chart. The mnemonic code is PER, and the octal code is -01 with optional index interval bits. It is a category B instruction with the *same* common commands as the *halt*. Review these on the common command chart.

The *one and only* unique command occurs at OTA-9. This is *104* which checks the operate gates and operates the *specified* equipment. The index interval bits will provide an *enabling* level for the proper gate. Check the index interval list to see which operations are available, and the bit configuration which specifies each.

Shift Left and Round

The *shift left and round* instruction shifts information from the B register into the accumulator and rounds the number off. Figure 12-6 is the timing chart.

This is a category A instruction which uses a common *PT cycle* and an arithmetic *pause*. The mnemonic code is SLR, and the octal code is 02. It uses *neither* index interval code nor data address. The data *address* portion of this instruction word specifies how many *shifts* are to be performed before the *rounding* off process.

This instruction was designed to prepare *double length* words for storage. Double length words are generated by multiply and divide instructions. These instructions also leave the most significant bits in the *B register*. Shifting left 15_{10} shifts discards the entire accumulator contents (except the sign, which does not change) and moves the LSB of the B register to the B register sign position. (The B register under these circumstances contains 16 bits of unsigned data). When this last bit of the B register is a *one*, we round off by *adding one* to the C(Acc). When it is *zero, no addition* is required.

The actions described are used after a multiply or divide instruction, but the SLR instruction can be used at *other times* and in other ways. The only difference in actual operation is the number of shifts performed. We may specify any number of shifts from 0 to 77_8. Any number *other than 0* will generate a *pause* which will terminate when the step counter reaches zero.

The common commands for the SLR are the same as those previously described for a PT cycle. At PT9, we generate unique *command 138* if the step counter contains other than zero. At PT10, 134 is generated under the same conditions. 138 sets the 2 MHz Sync flip flop to prepare for *timing* during the pause, and 134 sets the *pause* flip flop.

During the pause, all actions are controlled by *2 MHz* pulses except commands 73, 2, 5, 81, and 82. 73 *reduces* the C(step Ctr) by *one* each 0.5 μs.

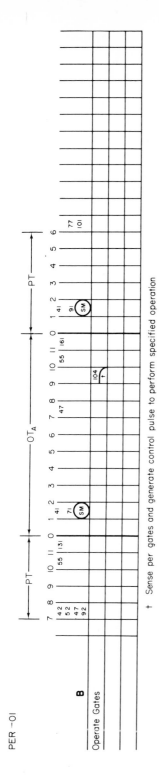

Fig. 12-5. Operate Timing Chart.

† Sense per gates and generate control pulse to perform specified operation

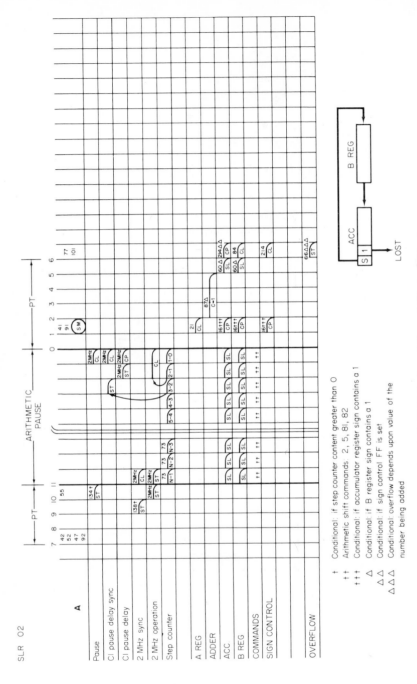

Fig. 12-6. Shift Left and Round Timing Chart.

2, 5, 81, and 82 work in combination to move the proper bits *one place* to the left. Each of these commands occur each 0.5 μs.

As the count in the step counter approaches zero, it signals the *end* of the pause. When the count changes from *2 to 1* a clear pulse is applied to the 2 MHz *operation* flip flop. This flip flop is cleared as the count changes from 1 to 0. The next pulse comes from the regular ring counter, and it is PT0.

If a *rounding* process is required, the instruction action will carry over into the PT cycle. While the *common* commands are bringing out the *next* instruction, the *unique* commands will *finish up* the SLR actions.

At PT1 we have commands 21 and 16. 21 *clears* the A register and 16 *complements* C(Acc), C(B register), and the sign control flip flop if the accumulator contains a *negative* number. At PT2, command 87 *adds one* to the accumulator contents by carrying a one to the LSB *adder*. At PT5, command 60 *shifts* our bits one place to the *left*. This is required after all additions because this machine has an *automatic right shift* inherent in the addition process. At PT6, we have 66, 84, and 214. 66 *checks overflow;* 84 *clears* the B register, *complements* the C(Acc), and *clears* the sign control flip flop. Nearly all of these commands depend upon specific *conditions* which are described on the timing chart.

Load B Register

The *load B register* instruction clears the B register and loads it with the word from a specified memory location. Figure 12-7 is the timing chart.

This is a category B instruction with an indexing option. The mnemonic code is LDB, and the octal code is −03 with a choice of index interval bits.

Shortly after PT8, we see common commands 114, 115, 212, 117, and 124. *Check the conditions* on the common command chart in Figure 12-2, and the *functions* on the common command list. You should have discovered that we can have *only one* of these commands, and its presence is *dependent* upon index interval bits other than 0000. In other words, if we *specify* an index register, its contents will be added to bits 10–15 of this instruction to *calculate* a new data address. Otherwise we take our data word from the memory location already specified by bits 10–15 of this instruction. All other common commands have been described previously.

The unique commands are few and simple; we have 84 at OTA6 and 39 at OTA7. 84 *clears* the B register, and 39 *moves* the data word from MIR into the B register.

Compare Full

The *compare full* instruction takes a word from the specified memory location and compares *each bit* with a corresponding bit in the accumulator.

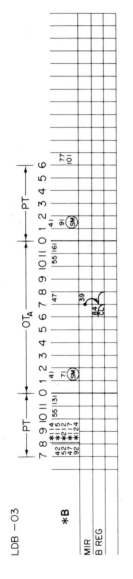

Fig. 12-7. Load B Register Timing Chart.

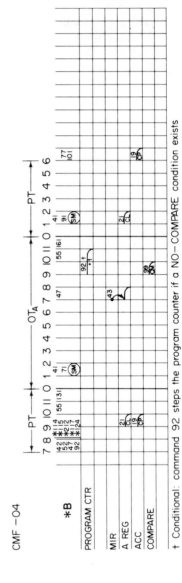

Fig. 12-8. Compare Full Timing Chart.

† Conditional: command 92 steps the program counter if a NO−COMPARE condition exists

338

If any two bits in corresponding positions are *different*, the program counter is caused to *skip* the next instruction. Figure 12-8 is the timing chart.

This is a category B instruction with an indexing option. The mnemonic code is CMF, and the octal code is −04 with optional index interval bits. The command commands have already been discussed.

At PT9, command 21 *clears* the A register, and 19 *complements* the C (Acc). Complementing the accumulator bits is necessary because of the nature of the compare *logic*. If we have two *identical* bits we want a *compare* condition. The comparator signals a *no compare* condition when the two bits are the *same*. This is *offset* by *complementing* the accumulator.

At OTA7, command 43 *moves* the word from the memory information register into the A register. At OTA9, command 99 makes a *bit by bit* comparison between the C(Acc) and C(A Reg). If one or more of the 16 bit positions generates a *no compare* signal, *one is added* to the program counter by command 92. Command 91 has already stepped the program counter as this instruction was extracted from memory. At this time, the program counter contains the address of the *next* sequential instruction. If we generate command 92, we change this address and cause the next sequential instruction to be *skipped*. The logic for this action is illustrated in Figure 12-9.

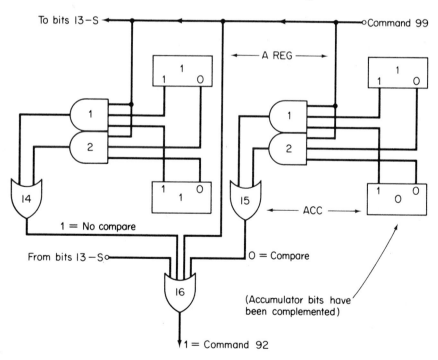

Fig. 12-9. Compare Logic.

If all corresponding bits were *identical* at the start, they will be *different* after complementing all accumulator bits. When command 99 is generated, *different bits* constitute a *compare* condition and *neither* AND gate has an output. This is shown in the bits 15 position. When both bits are *ones*, command 99 *passes* through AND 1. When both bits are *zeros*, command 99 *passes* through AND 2. An output from either AND gate activates the following OR gate and produces a *no compare* signal. A no compare signal from any (one or more) position will activate OR 16 and generate command 92. Notice that one leg of OR 16 is enabled by command 99. This causes command 92 to follow 99 very closely in case of a no compare.

Going back to our timing chart in Figure 12-8, we are just concluding the OTA cycle. The instruction is complete with the exception of a couple of *housekeeping* chores. At PT1, command 21 *clears* the A register. At PT6, command 19 *recomplements* the C(Acc) to restore it to the configuration that existed when this instruction started.

Test One Bit

The *test one bit* instruction causes a specified *bit* in a memory word to be *checked* and a *decision* rendered according to what the bit is. If the bit is a one, the *next* instruction will be *skipped*. Figure 12-10 is the timing chart.

This is a category B instruction. The mnemonic code is TOB and the octal code is −05. Index interval bits from 0000 − 1111 specify the *bit* to be tested.

At OTA1, command 102 is generated. It *sets* a flip flop to enable the gate of the *bit* to be tested. The next command is 43, which *moves* the selected word from MIR into the A register at OTA7. At OTA9, command 99 performs a *comparison* between the selected bit and the flip flop which was set by command 102. If the tested bit is a *zero*, the next instruction is performed in natural *sequence*. If the tested bit is a *one*, command 92 will be generated. In this case, one will be added to the program counter to increase the address by one. This causes the next instruction to be *skipped*.

Test Two Bits

The *test two bits* instruction enables the computer to examine *two consecutive bits* and make *four* decisions based on the bit configuration. The specified bit is tested *first*; then the bit to its *right* is tested. If bit 15 is specified, bit 15 and S are tested in that order. Figure 12-11 is the timing chart.

It is a category B instruction which uses index interval bits to specify the bits to be tested. The mnemonic code is TTB, and octal code is −06.

At OTA1, command 102 *sets* two flip flops to select the specified *bits*. At OTA7, timing becomes a bit crowded; commands 90, 99, and 43 appar-

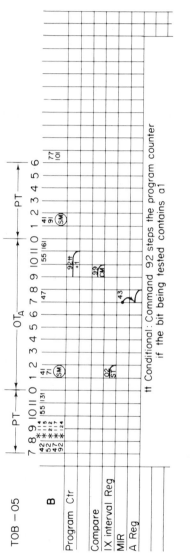

Fig. 12-10. Test One Bit Timing Chart.

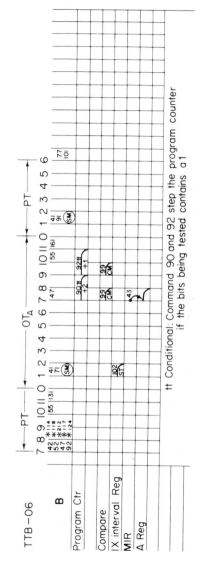

Fig. 12-11. Test Two Bits Timing Chart.

341

ently all occur at the same time. Actually 43 occurs first, then 99, and then 90, with very short delays in between.

43 *moves* the word from MIR into the A register; 99 performs the *comparison*; and 90 is conditional, depending on the *results* of the compare. At OTA9, command 99 occurs again to check the *second bit*, and command 92 is dependent upon this second result. If *both* bits are *zeros* neither 90 nor 92 will occur. In that case, the *next* instruction will be *executed*. If the specified bit is a *zero* and the one to its right is a *one*, command 90 does not occur but 92 does. This *adds one* to the program counter and causes the *next* instruction to be *skipped*. If the specified bit is a *one* and the one to its right is a *zero*, command 90 is generated, but 92 is not. This *adds two* to the program counter and causes the *next two* sequential instructions to be *skipped*. If *both* bits are *ones*, both 90 and 92 will be generated. This *adds three* to the program counter and causes the *next three* sequential instructions to be *skipped*. Reading the configuration of the tested bits as octal numbers summarizes the action.

SPECIFIED BIT	NEXT BIT	
0	0	SKIP NO INSTRUCTION
0	1	SKIP ONE INSTRUCTION
1	0	SKIP TWO INSTRUCTIONS
1	1	SKIP THREE INSTRUCTIONS

Clear and Add

The *clear and add* instruction is the most direct method of moving a word from memory into the accumulator. It causes the accumulator to be cleared, and moves the memory word into the cleared accumulator. This is accomplished by transferring the word from MIR to the A register and then *adding* C(A Reg) to a cleared accumulator. Figure 12-12 is the timing chart.

This is a category B instruction with an indexing option. The mnemonic code is CAD and the octal code is −10.

At OTA1, command 21 *clears* A register. Command 10 *clears* the accumulator at OTA6. At OTA7, command 43 *moves* the memory word from MIR into the A register. At OTA10, command 64 *carries a zero* to LSB adder to initiate an addition. The addition uses 1.5 μs, and at PT1, the memory word is in the accumulator. Commands 2, 4, 81, and 89 occur at PT1, command 21 clears the A register.

Add

The *add* instruction extracts the sum of a word from memory and a word in the accumulator. The sum of the two words replaces the original word in the accumulator. Figure 12-13 is the timing chart.

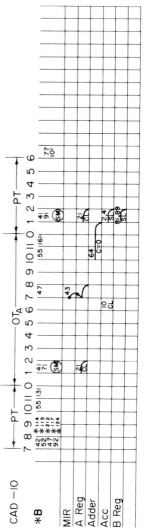

Fig. 12-12. Clear and Add Timing Chart.

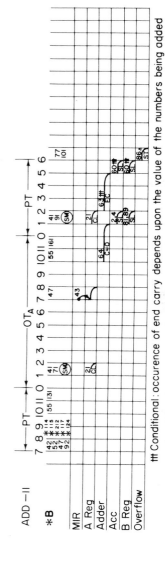

††† Conditional: occurence of end carry depends upon the value of the numbers being added

△ Conditional: occurence of overflow depends upon the value of the numbers being added

Fig. 12-13. Add Timing Chart.

343

This is a category B instruction with an indexing option. The mnemonic code is ADD, and the octal code is −11.

At OTA1, command 21 *clears* the A register. The memory word is *moved* from MIR to the A register at OTA7 by command 43. Addition *begins* at OTA10 when command 64 *carries a zero* to the LSB adder. The addition is *finished* by PT1 when 2, 4, 81, and 89 perform the *correctional* shift left and 21 *clears* the A register. This addition may or may not cause an *end carry*. If it does not, the action is finished by PT2, except for checking overflow. If end carry results, command 63 is generated at PT2. This *carries a one* to the LSB adder and initiates a *second* addition. Since the A register is clear, this second addition increases C(Acc) *by one*.

After the second addition is completed at PT5, command 60 will perform the *correctional* left shift. At PT6, command 66 checks for *overflow*, and if it is present, sets the overflow flip flop to *record* this condition.

Add B Register

The *add B register* instruction extracts the sum of the words in the B register and in the accumulator. The resulting sum replaces the original word in the accumulator. Figure 12-14 is the timing chart.

This is a category B instruction with no options, and it has *no need* for a *data address*. The bits S–3 and 10–15 have no meaning. The mnemonic code is ADB, and the octal code is 12.

The ADB is *identical* to the ADD with the *exception* of two commands. At OTA9, command 21 *clears* the A register. Command 88 *moves* C(B Reg) into the A register at OTA9. The remainder is a *normal* addition process.

Clear and Subtract

The *clear and subtract* instruction clears the accumulator and moves the complement of the selected memory word into the accumulator. The timing chart is shown in Figure 12-15.

This is a category B, indexable instruction. The mnemonic code is CSU, and the octal code is −13. This instruction is *identical* to the CAD instruction *except* for one command. Command 26 *complements* the A register at OTA9.

Subtract

The *subtract* instruction extracts the difference between a selected memory word and the word in the accumulator. The difference replaces the original C(Acc). Figure 12-16 is the timing chart.

This is a category B instruction with an indexing option. It is *identical* to ADD *except* for one command. At OTA9, command 26 *complements* the A register. The mnemonic code is SUB, and the octal code is −14.

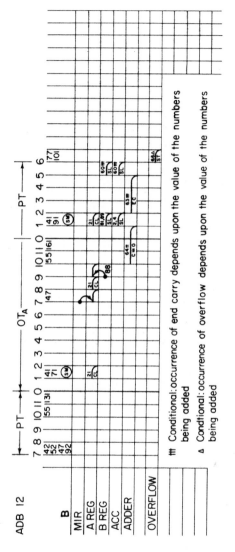

Fig. 12-14. Add B Register Timing Chart.

345

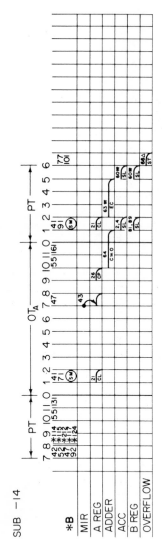

Fig. 12-15. Clear and Subtract Timing Chart.

††† Conditional: occurrence of end carry depends upon the value of the numbers being added

∆ Conditional: occurrence of overflow depends upon the value of the numbers being added

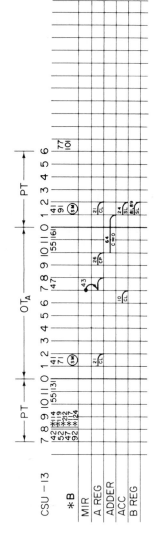

Fig. 12-16. Subtract Timing Chart.

Multiply

The *multiply* instruction extracts the product of a specified memory word and the word in the accumulator. The product is a double length word which occupies both the B register and the accumulator. The *most* significant bits will be in the *B register*. Figure 12-17 is the timing chart.

This is a category B instruction with an indexing option and an arithmetic pause. The mnemonic code is MUL, and the octal code is −25. The pause occurs after execution of the OTA cycle.

At OTA1, 21 *clears* the A register. Several things happen at OTA7: 43 *moves* the memory word into the A register; 75 *sets* a count of 15_{10} into the step counter; 13 *complements* the accumulator if necessary to make it *positive*. If the accumulator was negative, 13 also *complements* the sign control flip flop, and 84 *clears* the B register. At OTA8, 138 *sets* the 2 MHz Sync, 22 makes the A register *positive* (complements if necessary), 1 *moves* C(Acc) into the B register, and 10 *clears* the accumulator. If the A register was *negative*, command 22 also *complements* the sign control flip flop.

The 2 MHz operation becomes *effective* at OTA9, and 134 *sets* the pause flip flop at OTA10. The *first* arithmetic operations occur at OTA10 with commands 83 and 73. 83 *shifts* the B register one position to the right and checks B-15. If B-15 is a *one*, C(A Reg) is *added* to C(Acc) by *carrying a zero* to the LSB adder. This addition is accompanied by the *inherent* shift right. At the same time, command 73 *reduces* the step counter from 15 to 14.

Commands 83 and 73 are *repeated* every 0.5 μs through the pause and up to PT3. When the step counter changes from *7 to 6*, action is initiated to *restart* the normal timing. 1 μs later, the *pause* flip flop is *cleared*, and the ring counter *restarts* with PT0. Meanwhile, we have used *three* more 0.5 μs intervals, and multiplication is *continuing* under control of the 2 MHz operation.

As the step counter changes from *2 to 1*, the 2 MHz *operation* flip flop starts to clear. The final 73 and 83 occur at PT3. At PT6, command 9 *clears* the sign control flip flop, and if this flip flop was *set*, goes on to *complement* all bits in *both* the B register and the accumulator.

Divide

The *divide* instruction divides the combined C(Acc and B Reg) by a word from a selected memory location. The *quotient* will be in the B register, with the *remainder* in the accumulator. The timing chart is shown in Figure 12-18.

This is a category B instruction with an indexing option and an arithmetic pause. The mnemonic code is DVD, and the octal code is −26.

At OTA1, 21 *clears* the A register. At OTA7, 74 *sets* 17_{10} into the step counter; 43 *moves* the divisor into the A register; and 16 makes C(Acc and

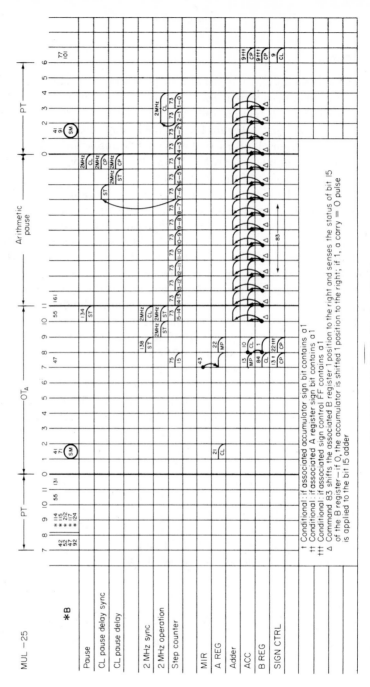

Fig. 12-17. Multiply Timing Chart.

† Conditional: if associated accumulator sign bit contains a 1
†† Conditional: if associated A register sign bit contains a 1
††† Conditional: if associated sign control FF contains a 1
△ Command 83 shifts the associated B register 1 position to the right and senses the status of bit 15 of the B register—if 0, the accumulator is shifted 1 position to the right; if 1, a carry = 0 pulse is applied to the bit 15 adder

348

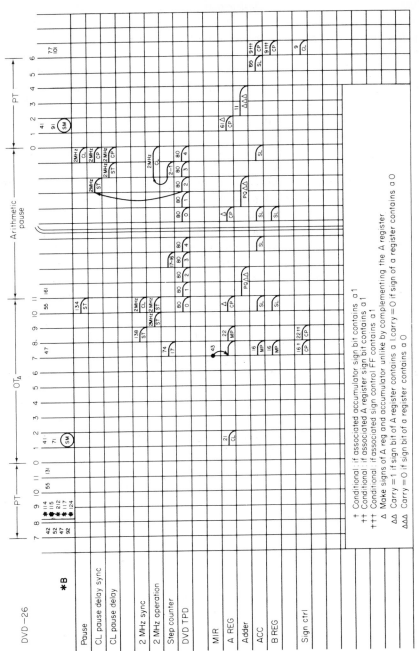

Fig. 12-18. Divide Timing Chart.

349

B Reg) *positive*. If this is a *negative* quantity, command 16 also *complements* the sign control flip flop.

At OTA8, 138 *sets* the 2 MHz Sync, and 22 *makes* the A register *positive*. If A register is *negative*, command 22 also *complements* the sign control flip flop. 2 MHz *operation* is now in effect.

At OTA10, 134 *sets* the pause flip flop, the A register receives a *conditional* complement, and the *dividend* is shifted left. The divide timing pulse distributor (TPD) is *started* and produces TP0. During each TPD cycle, *5 pulses* are generated, command 80 occurs *5 times*, the step counter is *reduced* by one, a *partial quotient* is generated, and the accumulator is *shifted* left. (This is the correctional shift). All of this work is accomplished by command 80 with some help from the TPD pulses.

A partial quotient is *produced* by adding C(A Reg) to C(Acc). This addition is initiated by *carrying the sign* of the A register to the LSB adder. 16_{10} partial quotients are required, so the 2 MHz *operation* is cleared when the step counter changes from *2 to 1*. Prior to each addition, the *signs* of accumulator and A register are made *unlike* by complementing the A register when *necessary*.

After the *16th* partial quotient, the *pause* terminates and *normal* timing *resumes* with PT0. At PT1, command 61 again makes the *signs* of accumulator and A register *unlike*. At PT2, we may have an *overdrawn* condition. This is determined by the A register sign. If the sign is *positive*, we *are* overdrawn. This means we must *replace* the last quantity that we subtracted. It is corrected by command 11, which *carries a zero* to the LSB adder. At PT5, command 60 *shifts* left if last described addition was performed. At PT6, command 9 *clears* the sign control flip flop, and if this flip flop is *set*, it also *complements* all bits in the accumulator and B register.

Remember that we are dealing with a *fractional* machine. When composing data for a divide instruction, we must be careful that the divisor is *larger* than the dividend. Otherwise our quotient will be a *whole number*, which is *too large* for the machine.

Store

The *store* instruction stores the accumulator word into the selected memory location. Figure 12-19 is the timing chart.

This is a category C instruction with an indexing option. The *address* portion of the instruction word *selects* the memory location for storage. The mnemonic code is STA, and the octal code is -34.

This instruction uses a PT and an OTB cycle. Only one unique command is required. At OTA2, command 17 *moves* the accumulator word into MIR. The *common* commands do the rest. The core location has been selected and a memory cycle started. Core contents are read out at OTA3 but IM (en-

Fig. 12-19. Store Timing Chart.

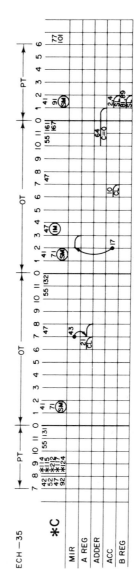

Fig. 12-20. Exchange Timing Chart.

351

circled) *prevents* the transfer from core to MIR. During the *write* portion of the same memory cycle, the contents of MIR (our word from the accumulator) are automatically *written* into the location which has just been *cleared*.

Exchange

The *exchange* instruction exchanges the C(Acc) with the C(selected memory location). Figure 12-20 is the timing chart.

This is a category C instruction with an indexing option. It uses all *three* machine cycles. The mnemonic code is ECH, and the octal code is -35.

At OTA6, command 21 *clears* the A register. The memory word is *moved* from MIR into the A register at OTA7 by command 43.

At OTB2, command 17 *moves* the accumulator word into MIR. *Normal* memory action will store this word into the core location.

At OTB6, command 10 *clears* the accumulator. The memory word is *moved* into the accumulator by an *addition*. The addition begins at OTB10 when command 64 *carries a zero* to the LSB adder. The addition is followed by a *correctional* shift left at PT1.

Shift Class Instructions

The *shift class instructions* allow us to *rearrange* data in a variety of ways by shifting. The timing chart is determined by the *number of shifts* specified. There are *six* instructions in the group. Figure 12-21 illustrates the timing and choice of shifts.

Basically the *address* portion of the instruction specifies the *number* of shifts. This number is set into the step counter. The proper transfer paths are completed and shifting begins. After each shift, command 73 *reduces* the step counter by *one*. When the count changes from *one to zero*, shifting ceases. Only a PT cycle is required for *1 to 6* shifts. *Over six* shifts require an arithmetic pause which terminates as the step counter changes from *7 to 6*.

Branch Positive Index

The *branch positive index* instruction can be used to branch *unconditionally* to a new section of the program. It can also branch on the *condition* that a specified index register contains a *positive* number. Figure 12-22 is the timing chart.

This is a category A instruction to branch program control. The mnemonic code is BPX, and the octal code is -51. The index interval bits are used to specify an *index* register. If these bits are 0000 or 0011, we have an *unconditional* branch. Any other bit configuration will select index register 1, 2, 4, or 5. In this case, we branch only if the *selected* register contains a

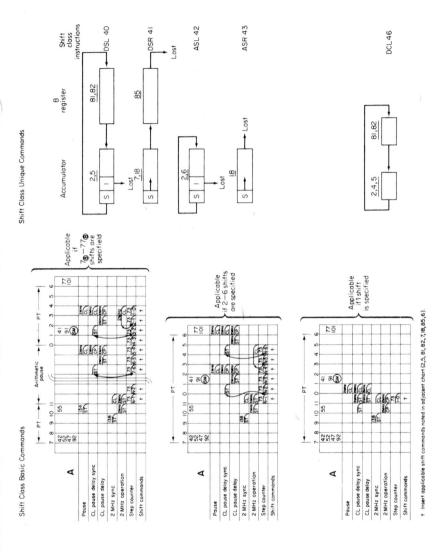

Fig. 12-21. Timing and Control for Shift Class Instructions.

† Insert applicable shift commands noted in adjacent chart (2,5, 81, 82, 7,18, 85,6).

353

BPX − 51

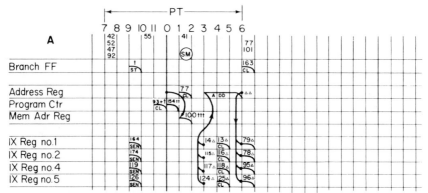

† Conditional: If IX0 or IX3 is selected, command 170 sets the branch FF to branch
  unconditionally. If IX1, IX2, IX4 or IX5 is selected, command 164, 174, 119 or
  126 sets the branch FF only if the selected index register has a positive − sign
†† Conditional: Branch FF must be set.
††† Conditional: Branch FF must be set. If branch FF is cleared, instruction address
   is transferred from program counter command 91.
 △ Conditional: Branch FF must be set and index register must be selected
△△ Conditional: Transfer from address register to selected index register can
   only occur if bit S of address register contains a 0

Fig. 12-22. Branch Positive Index Timing Chart.

positive number. The *address* portion of the BPX provides the address of the
next instruction if branching occurs.

At PT9, we *set* the branch flip flop if the conditions are met. If we do not
set the branch flip flop, *no* branching occurs, and no further unique commands will be developed.

Assuming a branching condition, command 93 *clears* the program counter
at PT11. Command 154 then *moves* the C (address register) into the program
counter, and 100 *moves* it into MAR. Command 77 *clears* the address register
at PT1. At PT3, the C(selected Ix Reg) is either *incremented* or *decremented*
by one. This is done through the address adder. The *result* of this addition will
be in the address register. If it is a *positive* quantity, it is *returned* to the index
register at PT6.

Branch Full Zero

The *branch full zero* instruction enables a branching decision to be based
on the contents of an accumulator word. It causes *all bits* to be examined,
and if all bits are *zeros*, branching will occur. Remember that a negative
number is stored in *complement* form. This means that we have *two* configurations which constitute zero. A number composed of *all zeros* is a *positive*

zero; a number of *all ones* is a *negative zero*. Either of the zero configurations will cause a branch. Figure 12-23 is the timing chart.

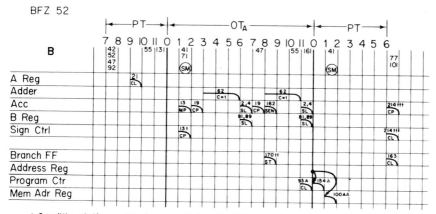

† Conditional: If associated accumulator sign bit contains a1
†† Conditional: Set branch FF if accumulator bit S ——— contains a1
††† Conditional: If associated sign control FF contains a1
△ Conditional: Branch FF must be set
△△ Conditional: Branch FF must be set. If branch FF is cleared, instruction address is transferred from program counter by command 91.

Fig. 12-23. Branch Full Zero Timing Chart.

This is a category B instruction. The mnemonic code is BFZ, and the octal code is 52.

At OTA1, command 13 makes the accumulator *positive*, and if it is neces- sary to complement the bits in the accumulator, it also *sets* the sign control flip flop. Now that we know the number is positive, command 19 *comple- ments* it at OTA2. Command 62 then *carries a one* to the LSB adder. Since the A register is clear, this addition *adds one* to C(Acc). After the addition, we *shift left* by commands 2, 4, 81, and 89, then *recomplement* the C(Acc). If the starting number was either a *positive* zero or a *negative* zero, this addition will cause *end carry* and leave the accumulator with a *positive sign*. When this is complemented, we have a negative sign. Command 162 checks the sign at OTA8, and if it is *negative*, enables a gate which produces command 170. 170 *sets* the branch flip flop, but before branching, we want to *restore* the accumulator to its original condition. The second command 62 does this by *carrying another one* to the LSB adder at OTA8.

Branching is accomplished by command 93 *clearing* the program counter at OTA11, followed by commands 154 and 100 at PT0. 154 *moves* the address portion of this BFZ instruction from the address register to the program counter. 100 *moves* this same address into the MAR.

At PT6, 163 *clears* the branch flip flop, and 214 *clears* the sign control flip flop. If the sign control is *set*, 214 also *complements* C(Acc).

Branch on Minus

The *branch on minus* instruction enables a branching decision based on the *sign* of the word in the accumulator. Branching occurs only if the sign is *negative*. Figure 12-24 is the timing chart.

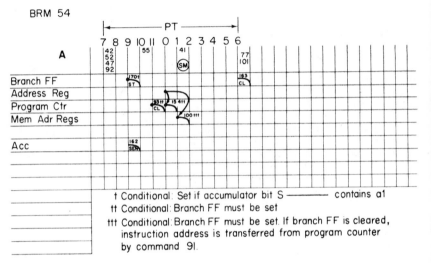

Fig. 12-24. Branch on Minus Timing Chart.

This is a category A instruction. The mnemonic code is BRM, and the octal code is 54.

It is a very simple instruction. At PT9, 162 *senses* the accumulator sign bit. If the bit is a *one*, command 170 is generated to *set* the branch flip flop. 93, 154, and 100 complete the branching action.

Load Counter

The *load counter* instruction is *one of three* instructions necessary to initiate an I/O operation. It loads the I/O address counter with the address portion of this instruction. This is the *starting* address in core for storing or extracting the information to be moved. Figure 12-25 is the timing chart.

This is a category A instruction with an indexing option. The mnemonic code is LDC, and the octal code is −60.

When the LDC is moved from the MIR, the *address* portion moves into the address register. At PT2, 148 *clears* the I/O address counter.

LDC-60

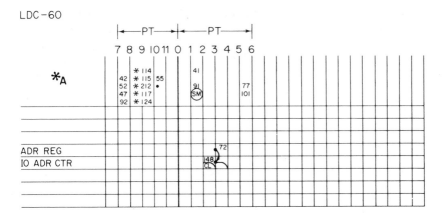

Fig. 12-25. Load Counter Timing Chart.

At PT3, command 72 moves the address from the address register into the I/O address counter.

When the transfer of information begins, the I/O address counter *replaces* the program counter. This address being loaded in just gives it the starting point.

Select Drum

The *select drum* instruction selects the drum as an I/O device for an I/O operation. Figure 12-26 is the timing chart.

This is a category B instruction with an indexing option. The *address* portion of the word is the starting address of the desired drum location where

SDR-61

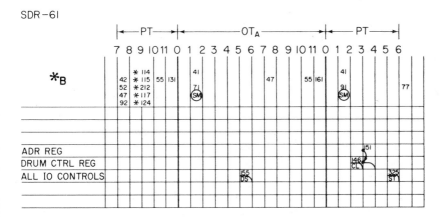

Fig. 12-26. Select Drum Timing Chart.

the information is to be either stored or removed. The mnemonic code is SDR, and the octal code is −61.

155 *deselects* (disables) I/O controls at OTA5. At PT2, 146 *clears* the drum control register. At PT3, 151 *moves* the starting data address from the address register into the drum control register. At PT5, command 325 *sets* the I/O controls (in this case, drum controls). The drum has been selected only. We are preparing for an I/O operation which will use the drum.

Select

The *select* instruction is used to select all I/O devices *except* drums. The index interval bits specify which device is selected. Figure 12-27 is the timing chart.

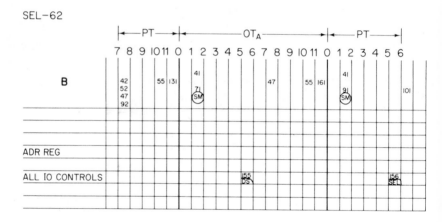

Fig. 12-27. Select Timing Chart.

This is a category B instruction. The mnemonic code is SEL, and the octal code is −62.

155 *deselects* the I/O controls at OTA5. At PT5, 156 *selects* the device specified by the index interval bits.

Read

The *read* instruction is used to move a specified *number* of words from the selected I/O device into core memory. Figure 12-28 is the timing chart.

This is a category A instruction with an index option. The mnemonic code is RDS with an octal code of −66. The device has already been selected by a previous instruction, but we may wish to *compute* the number of words to be transferred. If an index register is specified, the *number of words* is equal to C(Ix Reg) + C(Add Reg). If no index register is used, the address register

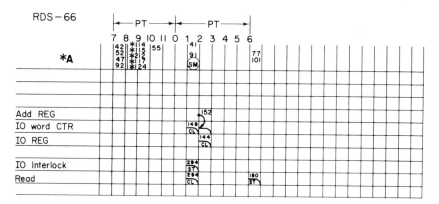

Fig. 12-28. Read Timing Chart.

contains the number of words (the address portion of this RDS instruction). 149 *clears* the I/O word counter while 294 *sets* the I/O interlock and *clears* the read controls. At PT2, 152 *moves* the C(Add Reg) into the I/O word counter. This number specifies the number of words to be moved. 144 *clears* the I/O at the same time. At PT6, 180 *starts* the reading action.

At this time, we *stop* the normal timing and *revert* to input/output timing. More about this after we examine the other instructions.

Write

The *write* instruction causes a specified number of words to transfer from core memory to the selected I/O device. The *address* portion of this instruction specifies the *number* of words to be transferred. Figure 12-29 is the timing chart.

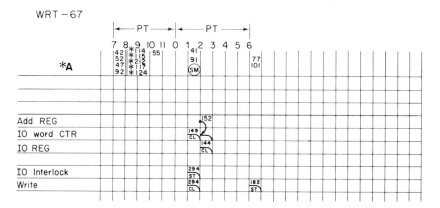

Fig. 12-29. Write Timing Chart.

This is a category A instruction with an indexing option. If indexing is used, we will *modify* our word count. The mnemonic code is WRT, and the octal code is −67.

At PT1, 149 *clears* the I/O word counter while 294 *clears* the I/O interlock and the write controls. At PT2, 152 *moves* the address portion of this WRT instruction (out word count) from the address register to the I/O word counter, and 144 *clears* the I/O register. At PT6, 182 *starts* the write operation.

Load Index with Address

This instruction causes the address portion of the instruction to be loaded into the specified index register. The index interval bits specify the index register to be selected. If these bits are 0000 or 0011, the instruction will be cycled through, but *no* useful work is done. Figure 12-30 is the timing chart.

This is a category A instruction. The mnemonic code is XIA, and the octal code is −75.

At PT4, the selected index register will be *cleared*. At PT6, the *address* portion of this XIA instruction is *transferred* from the address register into the cleared *index* register.

Load Index from Accumulator

This instruction *moves* accumulator bits S-6 into the selected *index* register. Index interval bits specify the index register to be *loaded*. Figure 12-31 is the timing chart.

This is a category A instruction. The mnemonic code is XAC, and the octal code is −76.

At PT9, 77 *clears* the address register which contains the *address* portion of this XAC instruction. At PT10, this is *replaced* as command 212 *moves* the seven high order bits from the accumulator into the address register.

At PT4, the selected index register is *cleared*. At PT5, the new number *moves* from the address register into the cleared index register.

A *negative* number *will not* load into an index register. Clearing an index register and moving a number from the index register leaves the sign bit *set*. This negative sign bit indicates an *empty* register, and it can be refilled *only* with a positive signed number.

Initiating an I/O Operation

Three instructions are necessary to *start* an I/O operation. These are LDC, SDR or SEL, and RDS or WRT. LDC *places* the starting memory address into the I/O address counter. SDR or SEL *selects* the peripheral device to be used. RDS or WRT establishes the *direction* of transfer and the *number* of words to be moved.

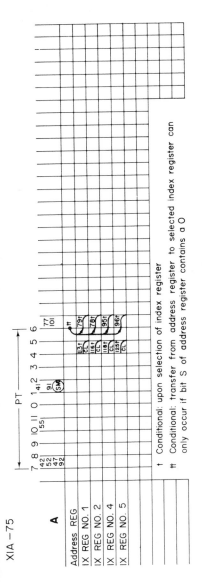

Fig. 12-30. Load Index with Address Timing Chart.

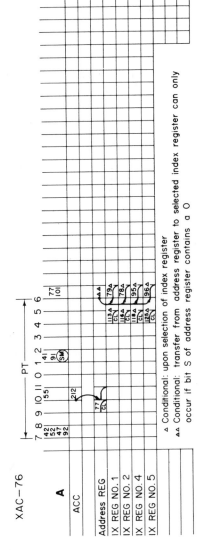

Fig. 12-31. Load Index from Accumulator Timing Chart.

361

Controlling the I/O Operation

After the I/O operation has been initiated, we set up a *transfer system* as shown in Figure 12-32.

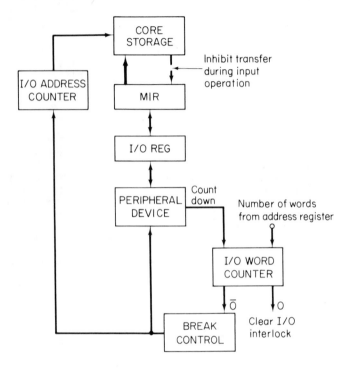

Fig. 12-32. I/O Control.

This transfer system *shares* core storage and MIR with the central processor. With an asynchronous system, the central processor *continues* to execute program instructions during the I/O operation. Several instructions can be executed in the time required for some peripheral devices to process one computer word.

When a word is *ready* to transfer, the central processor takes a 6 μs *break*. The break control takes over during breaks. The memory address is *selected* from the I/O address counter and this count is *increased* by *one*. A *normal* memory cycle then takes place in core and the MIR. On *input* operations, the location is *cleared* and replaced by a *new word* from the I/O register. On *output* operations, the memory word *transfers* to the I/O register. When the word transfer is complete, the I/O word count is *reduced* by one.

The control processor then *resumes* operation until another word is *ready* to transfer. A *6 μs break* and a normal *memory cycle* (either storage or retrieval) are used for *each word* transfer.

When the I/O word count reaches *zero, no* further breaks are requested, the I/O interlock is *cleared,* and the operation *ceases.*

Review Exercises

1. What is the transfer path from core storage to the accumulator?

2. What is the primary use of the address adders?

3. Name the functional component which normally supplies the memory address during:
   a. PT cycles.
   b. OT cycles.
   c. Breaks.

4. When does the address register furnish the address of the next instruction?

5. Which bits of an instruction word contain:
   a. Index interval bits?
   b. Instruction code?
   c. Data address?
   d. Conditions of a branch?
   e. Address being branched to?
   f. Type of shift?
   g. Number of shifts?

6. Why is a correctional left shift required after each addition?

7. Where is accumulator bit 15 after a shift right?

8. List five commands that occur during the first half of each PT cycle and state the function of each.

9. a. Under what condition does the HLT use a pause?
   b. What causes this condition to exist?

10. List the unique commands generated by PER.

11. What is meant by the term indexing?

12. This instruction word has just been decoded: 0.101001100011110.
   a. What is the instruction?
   b. What octal location is given?
   c. Which index register is specified?

13. Suppose that the specified index register of item 12 contains $+24_8$.
   a. What is the calculated data location?
   b. What number is reinserted into the index register?

14. How can we avoid address modification when using an indexable instruction?

15. Explain how overflow is possible when subtracting?

16. How many μs are required to execute the:
a. STA?
b. BPX?
c. MUL?
d. DVD?

17. How many times is command 73 generated during MUL?

18. How many times is command 80 generated during DVD?

19. We have $+16_8$ in Ix Reg 2. Write the binary configuration which will use this number for indexing, store C(Acc) into memory location 24, and return $+15_8$ to the index register.

20. Why is it necessary to have command 64 during execution of the exchange instruction?

21. This instruction is decoded: 0.000100110010000.
a. What will be accomplished?
b. How many μs will be used?

22. Before the execution of the instruction in item 21, C(Acc) is 0.0101010 10101010, and C(B Reg) is 1.101010101010101. What are the contents of both registers after execution?

23. The following instruction is decoded immediately after that in item 21: 0.000000010101010.
a. What bits will be stored?
b. What is the octal storage location?

24. Write the binary configuration for an instruction that will cause:
a. An unconditional branch to location 64.
b. A branch to location 16 only if Ix Reg 4 contains a positive number.

25. The following binary numbers are three consecutive I/O class instructions: 0.000110000000001, 0.100100010000000, 0.111110110100001.
a. Which peripheral device is selected?
b. What is the starting core address?
c. What type of I/O operation is indicated?
d. How many words will be transferred?

26. Suppose that the next instruction (after the three instructions in item 25) is HLT. Estimate the number of μs in the halt I/O pause and explain why.

chapter thirteen

programming

After instructions and data have been transcribed into an input device, the central computer can take over complete processing. However, the procedural *steps* that are to take place must be precisely defined in terms of *operations* that the system can perform. A definition of one step in a procedure is an *instruction*. A series of instructions which define all the steps in an entire procedure is a *program*. The act of assembling instructions and data to cause the computer to perform a given task is *programming*.

13-1 PROGRAM CODING

Many systems of programming language are used by the programmer in coding his program. Perhaps the simplest one makes use of the mnemonic code, and specifies location of instruction, and location of data.

Line of Code

When using the mnemonic code, each *line* of code defines one specific *step* for the computer to perform. For example, 06 CSU 30 constitutes a line of code. When changed to machine language, the CSU becomes the binary code for the *operation* section of the instruction word. The binary equivalent of 30_8 is placed in the *address* portion of the word. This instruction word is then stored into memory location 06.

The mnemonic method of coding a program is simple, straightforward, and easy to understand. Each line of code provides a complete instruction word and the core location where it is to be stored.

A very important part of the programmer's code sheet is the *annotation* column. This enables the programmer to enter brief *notes* after each coded instruction. These are primarily notes to himself as reminders of *why* each operation was performed. Figure 13-1 illustrates the use of this column.

INST. ADDRESS	OPERATION	DATA ADDRESS	ANNOTATIONS
21	CAD	50	Fetch normal wage
22	ADD	51	Add overtime
23	SUB	53	Subtract deductions
24	STA	60	File amount due

Fig. 13-1. Code Annotations.

This program appears to be making up a payroll. The *normal* wage of this individual is in location 50. In location 21 we have CAD 50 which places the normal wage into the accumulator. The instruction in location 22 is ADD 51, which takes the *overtime* from location 51 and adds it to the normal wage. The instruction in location 23, SUB 53, subtracts the *deductions* from the total wages due. The accumulator now contains the amount of this employee's paycheck. The instruction STA 60 from location 24 causes the *amount due* to be *stored* into location 60.

The brief notes on each line of code aid the programmer when he is *testing* his own program. More important, they explain the steps of the program to *anyone* who may use it in the future.

Locating Program and Data

Normally there is *no* specific area of memory set aside for either program or data. However, it is a common practice to *group* the instructions together and place them in ascending sequential locations. This provides for the automatic access operation of the program counter. Of course, instructions are *not always* executed in this exact sequence. The order may be *changed* by an unconditional branch or by a decision-making instruction.

The data is more flexible. There is actually no sound reason for grouping the data in most cases. A data word can be inserted into *any* available core location. The instruction which uses a data word tells the computer *where* to find that word.

Regardless of location, correct addresses of both instructions and data are of *vital* importance. The computer *does not* think; it does exactly what it is told. A word which is obtained from memory during program time will be interpreted as an *instruction*. A word which is obtained during operate time will be handled as *data*.

13-2 BASIC ROUTINES

In-Line

An in-line routine has the instructions located in the exact sequence that they are to be executed. Figure 13-2 contains an in-line routine for solving a simple arithmetic problem.

LOCATION	OPERATION	ADDRESS	COMMENTS
OO	CAD	O7	Puts "a" into the accumulator
O1	ADD	10	Forms a + b and leaves it in accumulator
O2	ADD	11	Forms a + b + c and leaves it in accumulator
O3	ADD	12	Forms a + b + c + d and leaves it in accumulator
O4	ADD	13	Forms a + b + c + d + e and leaves it in accumulator
O5	STA	14	Stores results of a + b + c + d + e in memory location l2
O6	HLT	OO	Stops the computer

Fig. 13-2. In-Line Routine.

Iterative

Much of the flexibility of a computer rests in its ability to *modify* instructions. The iterative routine generally makes use of *indexing* to modify the data address. This enables a few instructions to be performed repeatedly using a new set of data with each sequence. We sometimes refer to this type of routine as a *controlled* loop.

Suppose that we have a series of ten numbers to be added. Instead of using the ADD instruction ten times, we may use a routine similar to that in Figure 13-3.

LOCATION	OPERATION	ADDRESS	COMMENTS
OO	CAD	30*	Obtain num. 12_8 from memory
O1	(1)XAC	—	Load 12_8 into **IX** reg 1
O2	(1OO1)CAD	65	Obtain number in location 77
O3	ADD	40	Take sum of C(40) + C(77)
O4	STA	40	Store sum
O5	(1)BPX	02	Branch until **IX** reg 1 is zero
O6	—	—	

*(Location 30 contains 12 octal)

Fig. 13-3. Iteration Using Indexing.

This routine loads an octal 12 into index register 1. It then uses this number, decrementing by one after each use, to *modify* the data address of the CAD instruction. The whole routine causes the C(Men. Loc. 40 and 66 through 77) to be added together. The sum is stored in location 40.

The binary numbers in parenthesis are the index interval bits. (1) XAC causes Ix Reg 1 to be *loaded* from the accumulator. (1001) CAD *adds* the C(Ix Reg 1) to 65 to compute the data address. Since this instruction has a one as the first index interval bit, it will cause one to be subtracted from C(Ix Reg) each time we execute the CAD.

On the *first* pass, we add C(40) to C(77) and store the sum in 40. The BPX examines the sign bit of the index register and branches control back to the CAD instruction if the sign is positive.

On the *second* pass, we add C(40) to C(76) and store in 40. This *repeats* until the twelfth octal pass when C(Ix Reg 1) has been reduced to 1. We add 1 to 65 for data address 66, then we subtract 1 from C(Ix Reg). Like this:

$$
\begin{array}{ll}
0.000\ 001 & \text{FROM IX REG} \\
\underline{1.111\ 110} & \text{ONE'S COMP OF 1} \\
1.111\ 111 &
\end{array}
$$

The result of 1–1 is a negative *zero* which *will not* transfer to the index register. However, each time we removed the number from the index register, we left it with the sign bit set. After we have added C(40) to C(66) and stored in 40, the BPX finds that the index register contains a *negative* number. This time the branch *will not* occur; the program counter moves on to 06 to obtain the next sequential instruction.

Comparison

Suppose that we wish to compare the magnitude of three numbers and move the largest number to a particular location. The routine in Figure 13-4 will do this.

LOCATION	OPERATION	ADDRESS	COMMENTS
00	CAD	24	W in accumulator.
01	SUB	25	W − X in accumulator.
02	BRM	05	If W less than X, branch to location 05.
03	CAD	24	If W greater than X, put W into accumulator.
04	BPX	06	Branch control to location 06.
05	CAD	25	If W less than X, put X into accumulator.
06	STA	27	Stores the larger, W or X in memory location 27.
07	SUB	26	Subtracts Y from larger W or X.
10	BRM	12	If Y greater than W or X, branch to location 12.
11	BPX	14	Branches to location 14.
12	CAD	26	If Y greater than W or X, Y in accumulator.
13	STA	27	Stores Y in location 27, replacing W or X.
14	NEXT INSTRUCTION		
24	W		Initial data.
25	X		
26	Y		
27	−		Storage for largest number.

Fig. 13-4. Comparison Routine.

This program routine compares W, X, and Y (located in 24, 25, and 26 respectively), selects the *largest* number, and stores it in location 27.

Branching

The branching instructions provide us with a great variety of *decision* making conditions. Suppose that we have a group of four numbers, M, N, O, and P located in memory positions 50, 51, 52, and 53 respectively. These numbers may be all positive, all negative, or part positive and part negative. We wish to take the sum of the *negative* numbers and place it in location 54, and take the sum of the *positive* numbers and place it in location 55. This can easily be done by using our branch instructions, as illustrated in Figure 13-5.

LOCATION	OPERATION	ADDRESS	REMARKS
00	CAD	50	First number into accumulator.
01	BRM	04	Branch if minus, go on if positive.
02	STA	55	Store positive number in location 55.
03	BPX	05	Unconditional branch to location 05.
04	STA	54	Store negative number in location 54.
05	CAD	51	2nd number in accumulator.
06	BRM	12	Branch if minus.
07	ADD	55	Add positive number in location 55.
10	STA	55	Store positive sum in location 55.
11	BPX	14	Unconditional branch to location 14.
12	ADD	54	Add negative number in location 54.
13	STA	54	Store negative sum in location 54.
14	CAD	52	3rd number in accumulator.
15	BRM	21	Branch if minus.
16	ADD	55	Add positive number in location 55.
17	STA	55	Store positive sum in location 55.
20	BPX	23	Unconditional branch to location 23.
21	ADD	54	Add negative number in location 54.
22	STA	54	Store negative sum in location 54.
23	CAD	53	4th number in accumulator.
24	BRM	30	Branch if minus.
25	ADD	55	Add positive number in location 55.
26	STA	55	Store positive sum in location 55.
27	BPX	32	Unconditional branch to location 32.
30	ADD	54	Add negative number in location 54.
31	STA	54	Store negative sum in location 54.
32	HLT	00	Stops computer with sum of positive numbers in location 55, and sum of negative numbers in location 54.
50	M		Initial data.
51	N		
52	O		
53	P		
54	–		Storage for sum of negative numbers.
55	–		Storage for sum of positive numbers.

Fig. 13-5. Use of Branching.

13-3 DEVELOPING A PROGRAM

The process of program development is divided into *four* phases: problem analysis, organization, coding, and testing. As a general rule, problem analysis requires the services of a mathematician. Since the outcome of the first phase affects the organization, the problem analysis is best handled by a mathematician-programmer or a *team* which has these combined skills. Organization, coding, and testing are handled by computer programmers.

Problem Analysis

Before an efficient program can be developed to solve a problem, it is essential to:

1. State the problem.
2. Analyze all the facts.
3. Reduce the problem to a mathematical expression.

The mathematical expression should state the problem in its *simplest* form. Complex statements can be reduced through the process of numerical analysis. Computers are frequently used to reduce the labor of extensive numerical analysis.

Organization

Program organization is the process of *sequencing* the operations into an order which will simplify coding, minimize execution time, and conserve storage space. It is not always possible to attain all three objectives. It is frequently necessary to *compromise* between minimum time and maximum conservation of storage space. However, the most *efficient* program is the one which solves the problem with a minimum number of instructions and in a minimum time.

A program *flowchart* is a useful tool in program organization. It helps to keep the entire program in view and aids in developing the proper sequence of operations.

Coding

Working from the flowchart, it is a relatively simple matter to *code* the program. The result of this phase should be a properly sequenced, symbolically coded program that is ready for testing. The choice of using the mnemonic code or some other programming language is arbitrary and the programmer's choice.

Program Testing

A program with any degree of complexity seldom, if ever, operates satisfactorily on the *first* trial run. The more complex the program, the greater the possibility of errors. Errors occur in coding, in interpretation of machine functions, in punching cards and in machine operation. These errors are doubly expensive because the programmer's time and the computer's time are wasted until the errors are located and corrected. Errors must be kept to a minimum by checking and rechecking every step in the programming process before inserting the program into the computer.

The first step in checking the program is to have it *reviewed* by another programmer. The person who writes the program has a tendency to see what he *wants* to see when checking his own program. The reviewing programmer should make a flowchart from the program code listing and compare it with the flowchart used in coding the program.

13-4 FLOWCHARTING

The principal value of a flowchart lies in the fact that it shows a lot at a glance. It *graphically* represents organized procedures and data flow. The broad essentials and many details are readily apparent.

Flowchart Symbols

In flowcharting, symbols and words *support* one another for maximum clarity. Standard symbols enhance the graphic clarity of flowchart functions. These symbols are shown in Figure 13-6.

In a program flowchart, the emphasis is on computer decisions and processing. The programmer uses the flowchart to develop each step of his program. He starts with symbols representing *major* functions. As the program develops, he extracts large segments and describes them in *detail* on subsidiary flowcharts. The finished flowchart is his guide to coding the program.

Flowcharting Techniques

We will use a simple problem to illustrate the use of flowcharts.

Problem: Select the largest of three numbers.

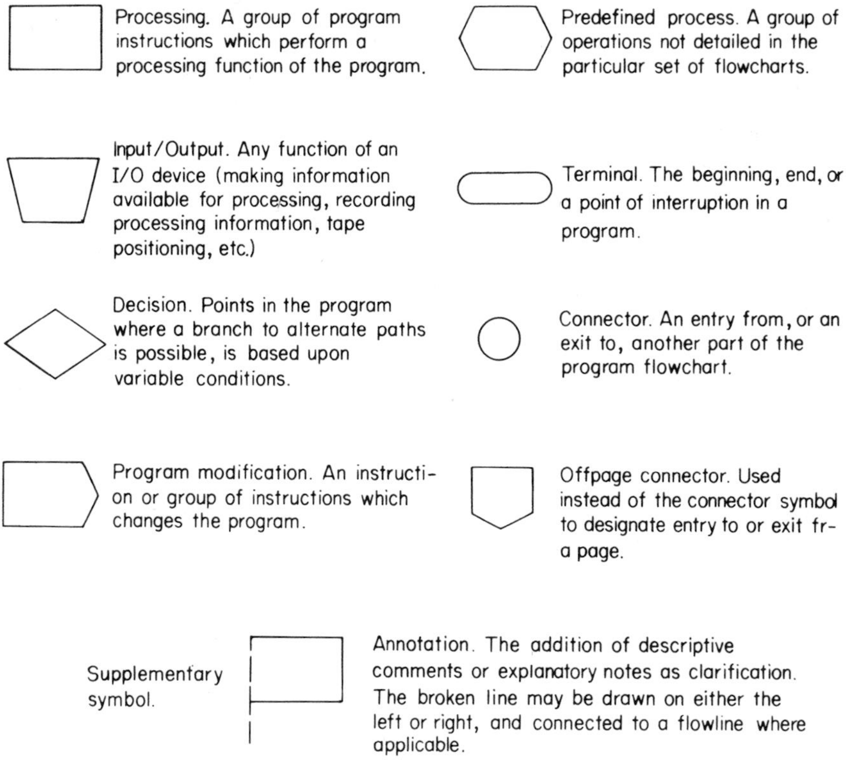

Processing. A group of program instructions which perform a processing function of the program.

Predefined process. A group of operations not detailed in the particular set of flowcharts.

Input/Output. Any function of an I/O device (making information available for processing, recording processing information, tape positioning, etc.)

Terminal. The beginning, end, or a point of interruption in a program.

Decision. Points in the program where a branch to alternate paths is possible, is based upon variable conditions.

Connector. An entry from, or an exit to, another part of the program flowchart.

Program modification. An instruction or group of instructions which changes the program.

Offpage connector. Used instead of the connector symbol to designate entry to or exit from a page.

Supplementary symbol.

Annotation. The addition of descriptive comments or explanatory notes as clarification. The broken line may be drawn on either the left or right, and connected to a flowline where applicable.

Fig. 13-6. Program Flowchart Symbols.

With this basic information, we can construct the preliminary chart in Figure 13-7.

A decision block follows each comparison because the larger number is not known. The programmer must follow *each* path from the decision blocks to its *ultimate* conclusion because he must provide for all selections and still select the largest number.

The chart seems to indicate that this process will take three *random* numbers and select the *largest* of the three. Now we add the code to perform the steps indicated on the chart. This produces the chart in Figure 13-8.

This is a *functional* program, and if used in this form, it will do the job. However, it can be made more *efficient*. Notice that in two cases, N_3 has been found to be the largest number. In each case, the decision is followed by *three* identical instructions. Let's combine those into a *single* sequence. When this is done, we still have three identical paths of CAD, STA, and HLT. We can

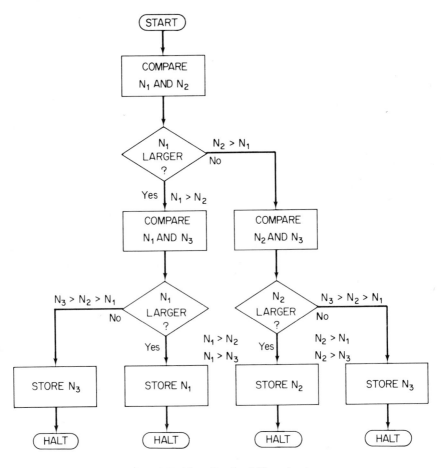

Fig. 13-7. First Draft of Flowchart.

use the unconditional branch to eliminate the *surplus* paths and *reduce* the number of instructions.

When we incorporate these changes into our flowchart, it becomes the final chart in Figure 13-9.

Now it is a simple matter to code the program from our *finalized* flow-chart. This is shown in Figure 13-10.

System Flowcharts

Computer systems designers prepare *general* flowcharts of the data system. Each chart shows all *related* processing steps. When these charts are

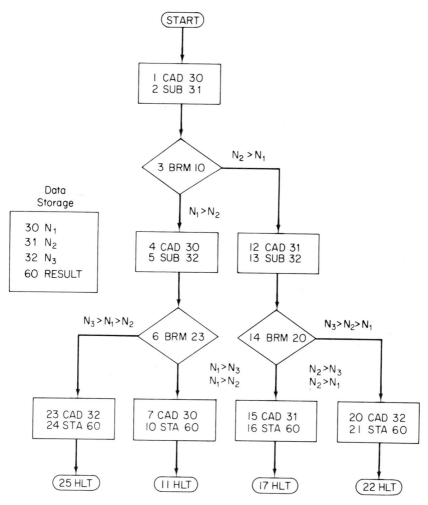

Fig. 13-8. Preliminary Flowchart with Code.

placed together, they provide a graphic representation of the *complete* system. They show all the *major* processing steps and tell *what* they are.

Information from the general flowcharts is *expanded* into detailed flowcharts. These detailed charts show *how* each major processing step in the general chart is accomplished.

The symbols for system flowcharting overlap into program flowchart symbols. In fact, a detailed *program* flowchart frequently uses several of the symbols from *system* flowcharts. Some of these symbols are illustrated in Figure 13-11.

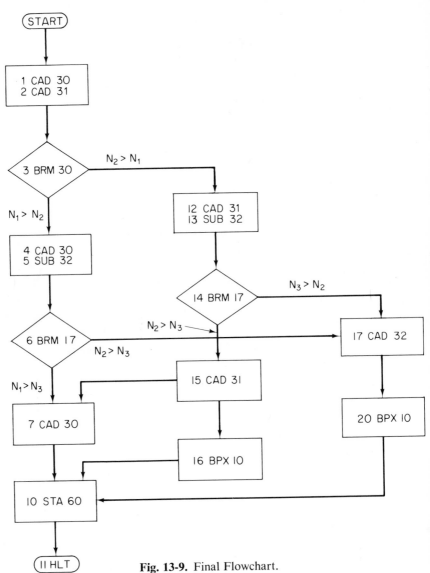

Fig. 13-9. Final Flowchart.

Location	Instruction Operation	Address	Comments	Contents of Accummulator
1	CAD	30	Places N_1 in accumulator	N_1
2	SUB	31	Subtracts N_2 from N_1	$N_1 - N_2$
3	BRM	12	Examines contents of accumulator if negative, branches program to instruction 12; if positive, goes on to instruction 4.	
4	CAD	30	Places N_1 in accummulator	N_1
5	SUB	32	Subtracts N_3 from N_1	$N_1 - N_3$
6	BRM	17	Examines contents of accumulator; if negative, branches program to instruction 17; if positive, goes to instruction 7.	
7	CAD	30	Places N_1 in accummulator	N_1
10	STA	60	Stores largest number in memory location 60	N_1
11	HLT		Stops computer operation	N_1
12	CAD	31	Places N_2 in accumulator	N_2
13	SUB	32	Subtracts N_3 from N_2	$N_2 - N_3$
14	BRM	17	Examines contents of accummulator; if negative branches program to instruction 17; if positive, goes on to instruction 15.	$N_2 - N_3$
15	CAD	31	Places N_2 in accummulator	N_2
16	BPX	10	Branches program to instruction 10	N_2
17	CAD	32	Places N_3 in accummulator	N_3
20	BPX	10	Branches program to instruction 10.	N_3

Data Storage Location	Contents Program	Comments
30	N_1	
31	N_2	
32	N_3	
60	—	Storage for largest number.

Fig. 13-10. The Coded Program.

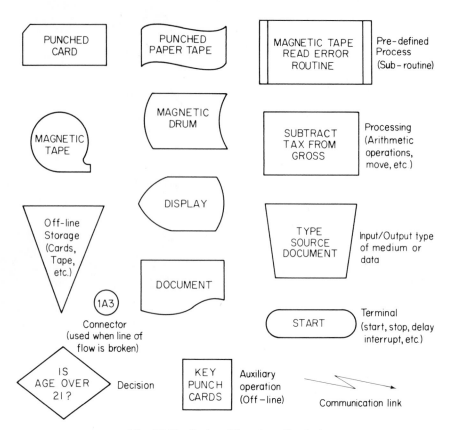

Fig. 13-11. System Flowchart Symbols.

Figure 13-12 illustrates a general and a detailed system flowchart.

Figure 13-12a is a *general* flowchart and 13-12b is a *detailed* flowchart. The entire chart of Figure 13-12b is used to explain the *computer* square on the general chart.

Notice that these charts contain *no* arrows on the flow lines. The rules of flowcharting are: start at the top left and flow *down* and to the *right*. When these rules are followed, no arrows are required. A chart which *deviates* from these conventions must use *arrows* to avoid confusion.

The flowchart must always indicate the *direction* in which events flow, but the flow lines must never cross. Yet, it is not always possible to connect the symbols with continuous flow lines. The *connector* symbol is used to show where interrupted lines are *joined*. This symbol is *never* used singly. One labeled connector shows where a line breaks or branches; other symbols with the *same* label show where the line is reconnected. Like an electric ground, all connector symbols bearing *identical* labels are at the *same* data flow point.

Notice that the detailed flowchart (Figure 13-12b) is, in effect, a program flowchart. This one is an input/output operation to read a deck of cards and to print the information from the cards to a hard copy document. It would be a simple matter to code this chart with the required instructions.

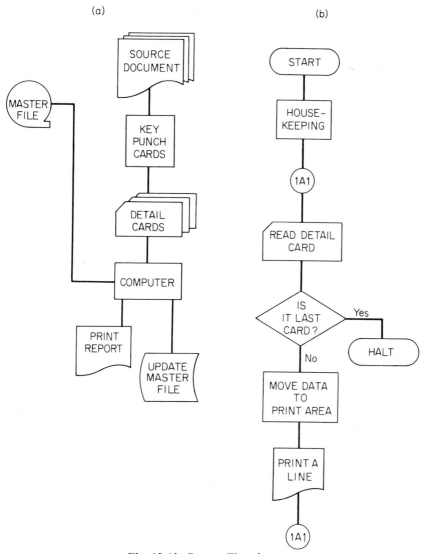

Fig. 13-12. System Flowcharts.

13-5 METHODS OF CODING

Machine Language

We know that the *language* of the central *computer* is a language of *binary* digits. Still dealing with our 16 digit computer, the instruction CAD 40 becomes 0.000001000100000. The machine code 0.101101001101110 tells the computer to branch to the instruction in location 56 if the content of index register 5 is a positive number.

This language of numbers is the *only* language the computer comprehends, and it is a *difficult* language for most of us. When writing or translating a number with many digits we are prone to error.

Symbolic Language

Several *symbolic* languages have been developed to bridge the gap between the spoken word and the machine language. Each of these has its own set of code and specifically defined symbols. Some examples are Programming Language 1 (PL/1), COBOL, and FORTRAN. COBOL uses business language, FORTRAN is a mathematical language, and PL/1 uses both.

COBOL is an *acronym* for the term "Common Business Oriented Language." FORTRAN was formed by *shortening* the term "Formula Translation." Still another symbolic language is the *mnemonic* code that we have used previously.

Regardless of the type of symbolic language used, *symbols* and abbreviated statements are used to *represent* operations. These operations will be *translated* into machine code.

13-6 TYPES OF PROGRAMMING

Two terms frequently encountered in computer programming are *micro* and *macro*. Micro programming refers to a *detailed* coding of minute operations. Macro programming is a technique of calling up *large* sequences of operations.

Micro Programming

Perhaps the greatest benefit of micro programming is to increase the *ability* of a *small* computer system. This is best described by an illustration. Suppose that we have a small system which has arithmetic functions of *only* add and subtract. If we wish to perform multiplication, we must program *each step* of addition, comparing, storing, etc. Instead of one instruction, MUL, we have a whole *routine*. The results are the same, but in *micro* programming, the effort rests with the programmer. Such a micro routine is illustrated in Figure 13-13.

(a) Subroutine for Multiplication

(b) Flowchart

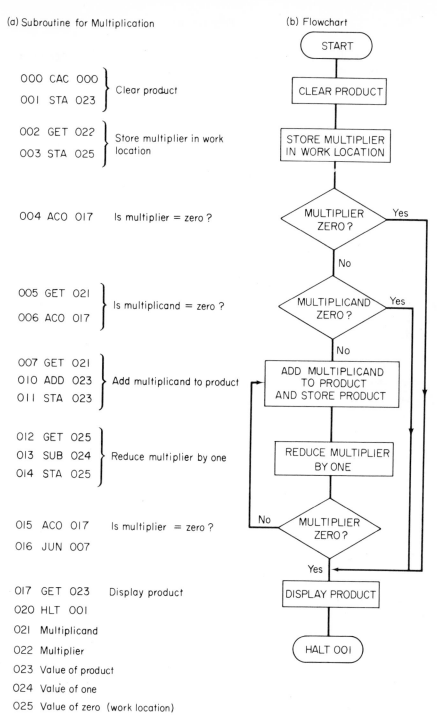

000 CAC 000	Clear product
001 STA 023	

002 GET 022	Store multiplier in work
003 STA 025	location

004 ACO 017   Is multiplier = zero ?

005 GET 021	Is multiplicand = zero ?
006 ACO 017	

007 GET 021	
010 ADD 023	Add multiplicand to product
011 STA 023	

012 GET 025	
013 SUB 024	Reduce multiplier by one
014 STA 025	

015 ACO 017   Is multiplier = zero ?

016 JUN 007

017 GET 023   Display product

020 HLT 001

021 Multiplicand

022 Multiplier

023 Value of product

024 Value of one

025 Value of zero (work location)

Fig. 13-13. Micro Program for Multiplication.

381

We are dealing with a different computer now, which brings up a few mnemonic codes that we have not previously discussed. The meanings of the new codes are:

CAC—Clear the accumulator.

GET—Same as CAD.

ACO—Branch to indicated address if C(Acc) = 0. If not continue to next instruction.

JUN—Branch unconditionally to indicated address.

Macro Programming

All large machines have *built-in* routines which we call *hardware* routines. The programmer makes full use of these automatic operations by *macro* programming. A single instruction may be used to call-up either a hardware, or a stored software routine of many steps. One routine may carry on a complete I/O operation; another may run a diagnostic test on several pieces of hardware. For a simple illustration of this process, let's go back to Figure 13-13.

This simple program could be used in a complex machine to perform multiplication. However, the complex machine would recognize the multiply instruction. By using CAD 021 and MUL 22, we can cause the machine to perform the whole routine. *Macro* programming is this same idea carried out to maximum efficiency.

13-7 TYPES OF PROGRAMS

Programs are designed to serve particular functions and are classified according to the type of function which they perform. All programs may be classified into one of *two* categories: *control* programs or *processing* programs.

Control programs *service* the execution of the processing programs; control the location, storage, and retrieval of data; and schedule jobs for continuous processing.

Processing programs *define* the work that the computer is to perform, and guide the computer through each step of performance.

Master Program

The *master* program is a *control* program. It does no processing on its own, but it calls up the various processing programs to perform the necessary data processing in the proper order. Other names for this control program are *sequence selection* program, *object* program, and *executive* program.

Utility Programs

Utility programs perform *nonchecking* functions such as *loading* other programs and *assembling* and translating symbolic programs. When the programmer finishes his symbolic program, he feeds it into a computer which is using a *utility* program. The utility program *translates* the symbols into machine language, composes addresses, and critiques the program. This type of utility program is frequently called either an *assembly* program or a *translator* program. Such programs can be written to understand and translate *any* given symbolic language.

Subroutines

Subroutines are *small* programs; each one has a relatively small task to perform. One subroutine performs an I/O operation; another may calculate a square root table. Many common routines may be available in various types of storage. The programmer makes full use of these by calling them up when his program needs their particular service.

Testing Programs

There are many types of *testing* programs. Some *exercise* the machine to prove its reliability; others *diagnose* malfunctions by testing specified sections of the hardware. There are even whole sets of test programs for testing new programs. The latter type of test program is especially helpful to the programmer. It can save him many hours of hard work and frustration when he is trying to learn why his program failed.

Review Exercises

1. Define the following terms as they apply to computer operations:
   a. Instruction.
   b. Program.
   c. Programming.

2. A line of code should contain three items of information for the machine and one item for the programmer. Describe these four items.

3. Figure 13-14 is a coded program. What are the C(Acc) after each step of the program?

4. Draw a flowchart for the program in Figure 13-14 and code it with the proper sequence of instructions.

5. Why do we normally group instructions and place them in a given order?

Location	Instruction Operation	Address	Comments	Contents of Accumulator
1	CAD	102	Places X in accumulator	
2	ADD	103	Adds H to X	
3	MUL	101	Multiplies (X + H) by P	
4	STA	700	Places P(X + H) in location 700	
5	CAD	100	Places C in accumulator	
6	SUB	700	Subracts P(X + H) from C	
7	STA	700	Places final solution in location 700	
10	HLT	——	Stops computer operation	

Data Storage Location	Contents	Comments
100	C	Capacity of fuel tank
101	P	Consumption rate
102	X	Distance flown
103	H	Altitude factor
700	—	Initial and final result storage

Fig. 13-14. Coded Program.

6. What type of instructions can change the order of instruction execution?

7. Why is it *not* necessary to place data in a given order?

8. What advantage is gained by using indexing in an iterative operation?

9. Name the four phases of program development.

10. Which of the development phases may be performed by a mathematician rather than a computer programmer?

11. What is the expected product of the problem analysis phase?

12. What are the characteristics of a program of maximum efficiency?

13. Which phases of program development are aided by use of a flowchart?

14. Why must each output from a decision block be followed to its ultimate conclusion?

15. Differentiate between a system flowchart and a program flowchart.

16. Differentiate between micro programming and macro programming.

17. Which of the following described systems would have the greatest requirement for micro programming:
    a. A system with a 16 bit word and a 4096 word memory?
    b. A system with a 49 bit word and a 24,536 word memory?

18. Why do we code programs in a symbolic language instead of writing directly in machine code?

19. COBOL is an acronym.
   a. It was coined from what words?
   b. What is COBOL?

20. Describe the process of translating a symbolic coded program into a machine language program.

21. Explain why every effort should be made to remove all errors from a program prior to trying a machine run.

appendices

BIBLIOGRAPHY

1. **Chu, Yaohan.** *Introduction to Computer Organization.* Prentice-Hall, NJ, 1970.

2. **Desmonde, William E.** *Computers and Their Uses.* Prentice-Hall, NJ, 1964.

3. **Fahnestock, James D.** *Computers and How They Work.* Ziff Davis, NY, 1959.

4. **Flores, Ivan.** *Computer Design.* Prentice-Hall, NJ, 1967.

5. **Flores, Ivan.** *Computer Logic.* Prentice-Hall, NJ, 1960.

6. **Jacobowitz, Henry.** *Electronic Computers.* Doubleday and Company, NY, 1963.

7. **Litton Industries.** *Digital Computer Fundamentals.* Prentice-Hall, NJ, 1965.

387

8. **Maley, Gerald & Skiko, Edward B.** *Modern Digital Computers.* Prentice-Hall, NJ, 1964.
9. **Mandl, Matthew.** *Fundamentals of Digital Computers.* Prentice-Hall, NJ, 1967.
10. **Robinson, Vester.** *Computer Concepts (a programmed manual).* Reston, Reston, Va, 1973.
11. **Sammet, Jean E.** *Programming Languages.* Prentice-Hall, NJ, 1969.

ANSWERS TO REVIEW EXERCISES

Chapter 1

1. Literally it means a finger or toe. It also means a single number.
2. It comes from a Latin word meaning pebble, which tells us that early man used pebbles as an aid to counting.
3. The abacus does not calculate and it does not make decisions. It is a mechanical aid to calculations.
4. It is a machine which can solve problems and make logical decisions.
5. They simplified the process of multiplication.
6. The machine parts lacked the necessary precision.
7. It was used on a weaver's loom.
8. The difference engine was intended to compute mathematical tables, and the analytical engine was designed to analyze mathematical formulas.
9. The perforated card concept and the Hollerith code.
10. The analog computer deals with physical quantities and repeatedly solves the same problem. The digital computer uses numbers and solves any problem that can be expressed in numbers.
11. A group of digits coded to cause the computer to perform a specific function.
12. It has created a great many new jobs.
13. A program is a list of instructions arranged in the proper sequence for accomplishing a specific task.
14. Input—transfers information from the operator to memory.
    Memory—stores information.
    Arithmetic—performs calculations and makes decisions.
    Output—transfers information from memory to the operator.
    Control—provides timing and control to guide functions of all other sections.
15. Program and data.

16. The input device translates the digits into electric pulses.

17. By a pulse.

18. By the absence of a pulse.

19. It also contains the location of the data needed to perform the operation.

20. Data are all facts and figures except program information.

21. Separate lines are sometimes used. If common lines are used, a timing sequence determines the nature of the information.

22. Digital and analog.

23. Military, commerce, industry, and science.

24. Not enough computers and prohibitive cost.

25. The student has individual attention; he can progress at his own rate; he has a greater variety of courses to choose from; and there is a uniform quality control on the instruction received.

Chapter 2

1. Both are data processors; both are programmed to follow a set procedure; the results are the same, but the computer is faster.

2. The design phase.

3. A careful analysis of the job the computer is expected to perform.

4. The control section.

5. The computer program.

6. A series of related storage devices for storing a complete computer word.

7. A coded instruction and the address of the data to be used with that instruction.

8. They are generally stored in sequential locations in the order they are to be used.

9. He manually inserts the address of that instruction.

10. By providing it with improper instructions or improper data.

11.

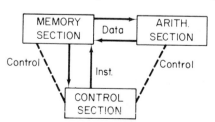

Fig. A-1. Control and Movement in Central Computer.

12. Addressable memory storage registers.

13. The contents of memory location 500 are 2000.

14. It holds the address of the next instruction and increments this address by one as each instruction is obtained.

15. Clear and add (CAD). The data moves from memory into the A register; the old data is cleared from the accumulator; and C(A register) is added to the zero in the accumulator. After the addition, the sum which is identical to the contents of the memory register will be in the accumulator.

16. The store (STR) instruction. The data are moved from the accumulator to the memory information register, and from that point into the specified memory storage register.

17. The location is specified by the address portion of the instruction word. This address is moved into the memory address register to select the location.

18. An address is selected, and the information in that location is moved into the MIR. From here it is rewritten into the same memory location and is gated out to either the arithmetic section or an output device.

19. As explained in item 18, each retrieve cycle rewrites the word into the same location it was taken from.

20. The master clock.

21. Clock pulses are a continuous string of pulses from the clock. Timing pulses are clock pulses within a cycle as they are grouped by the timing counter.

22. Program cycle—an instruction is removed from memory and decoded. Operate cycle A—Data is removed from memory and operated upon. Operate cycle B—Data is stored into memory.

23. It saves time by permitting simultaneous operations. This makes the information available precisely when it is needed in all cases.

Chapter 3

1. The language of numbers.

2. It must have a radix and follow a logical progression.

3. a. Two.　　b. Five.　　c. Eight.　　d. Sixteen.

4. a. 10.　　b. 16.　　c. 32.

5. a. 2^3 2^2 2^1 2^0 . 2^{-1} 2^{-2} 　　b. 8^3 8^2 8^1 8^0 . 8^{-1}
　　　 1　0　1　0 . 1　 1　　　　　 2　 0　 6　 4 . 2

c. 10^1 10^0 . 10^{-1} 10^{-2} 10^{-3} d. 16^2 16^1 16^0 . 16^{-1} 16^{-2}
   9   2 .   4   0   7   A   9   B . 4   C

6. a. The number is multiplied by 10.
   b. The number is divided by 10.

7. a. The number is divided by two.
   b. The number is multiplied by two.

8. The two symbols can be represented with a minimum of circuits.

9. a. Switch off.   b. Low voltage level.
   c. Absence of a hole in the tape.   d. Transistor cut off.

10. 0.0009765625.

11. Four.

12. 11.625_{10}.

13. a. 2^5.   b. 2^{-1}.   c. 2^{12}.

14. a. 110011111.001111_2.   b. 1000000000000.1001_2.
    c. 10001101010.000111_2.

15. It simplifies the conversion between binary and decimal.

16. 000011101010110_2.

17. a. 4096_{10}.   b. 0.000244140625_{10}.

18. 2418.58203125_{10}.

19. 46005.752225_8.

20. 101011100111_2.

21. 1151.224_8.

22. Binary, binary coded decimal, or hexadecimal.

23. Two.

24. a. 1111 0000 0000 1101_2, 170015_8, 61453_{10}.
   b. 1010 1011 1100 1101 1110 1111_2, 52746757_8, 11259375_{10}.
   c. 1110 1010 1100 1101_2, 165315_8, 60109_{10}.
   d. 0001 1000 1110 1011_2, 14353_8, 6379_{10}.

25. a. 2516254_8, 1010 1001 1100 1010 1100_2, $A9CAC_{16}$.
   b. 157257_8, 1101 1110 1010 1111_2, $DEAF_{16}$.

Chapter 4

1. a. 0000101010111100.   b. 1111101100001010.
   c. 0000011110100010.   d. 1111111111000000.
   e. 0111100101010011.   f. 1111111111111111.
   g. 0000000000010101.   h. 1111111111101010.

2. a. F134.   b. 110E.   c. 2EE35A.   d. ABCD.

3. a. 15047.   b. 111210.   c. 150535.

4. a. 65656.   b. 6364.   c. 15130.

5. a. 25255.   t. 22162.   c. 00726.

6. a. 1567520.   b. 2142044.   c. 2022377400.

7. a. 74.75.   b. 277.30.   c. 50.17.

8. a. 11000111.   b. 100101.   c. 10000100.

9. a. 101010.   b. 10110.   c. 110110.

10. a. 0110111.   b. 011111.   c. 0011011.

11. a. 1011101.   b. 110111.   c. 11101110.

12. a. 111.10.   b. 1000.00.   c. 1101.00.

13. a. 0000000000100111.   b. 1111111100100011.
    c. 1111111111110110.   d. 0000001000011110.

Chapter 5

1. Data signals and control signals.

2. By electric pulses. Usually the presence of a pulse is a one and the absence of a pulse is a zero.

3.

Fig. A-2. Wave Train of Binary Numbers.

4. When the bits are in temporary storage. A bit stored in a flip flop produces a constant output level for the duration of the storage.

5. They are the same circuits called by different names.

6.

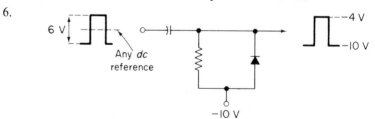

Fig. A-3. Biased Positive Clamper.

7. a. Unbiased negative clamper.
   b.

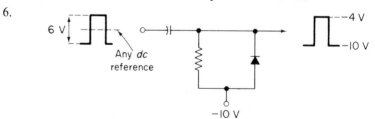

Fig. A-4. Output of Fig. 5-31.

8.

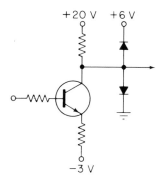

Fig. A-5. Establishing Output Limits.

9. a. This is a Biased Negative Limiter.
   b.

Fig. A-6. Output of Fig. 5-33.

10. To temporarily store binary bits.

11. Set, clear, and trigger.

12. a. Set.　　　 b. Clear.　　　 c. Trigger.　　　 d. Trigger.
    e. Clear.

13.

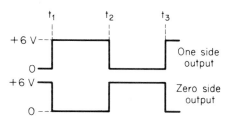

Fig. A-7. Outputs of a Flip Flop.

14. a. 2 MHz.　　　 b. 0.5 μs.

15. a. They guide the complementing trigger pulse to the cut off side of the flip flop.
    b. It increases the switching speed.

16. Common emitter.

17. a. Common base.　　　 b. Common collector　　 c. Common emitter.

18. To provide a 180° phase shift between input and output.

19. Basically all are the same. Buffer and driver are two names for the same amplifier. Amplifiers are named according to how they are used.

20. To synchronize all actions.

21. Any high frequency oscillator with a stable frequency used for computer timing. It is usually either crystal or tuning fork controlled.

22. The master clock can drive a limited number of circuits. When this number is exceeded, additional clocks are required.

23. The slave clocks have a free run frequency only slightly less than the master clock. This is to enable all clock frequencies to be synchronized with the frequency of the master clock.

24. 4 MHz.

25. 15 feet.

26. 4.22 MHz.

27. C_1 and R_1.

28. It provides an output when all the specified conditions are present.

29.

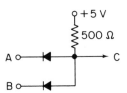

Fig. A-8. Two Legged Positive AND.

30. A positive potential equal to the level of a binary one.

31. a. A DTL OR Gate.
    b. A single positive pulse at either A, B, or C.
    c. A positive
    d. A zero.

32.

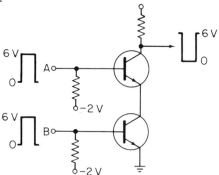

Fig. A-9. RTL NAND Gate With Wave Shapes.

33. All inputs must be at a zero level.

34. The XOR is a gate with two inputs. It has a high output when the two inputs are different.

35. Wave shaping. It provides a standard signal out as long as it has an activating level in.

Chapter 6

1. Both are widely used, and no standard has been established.
2. No. The symbol represents a function. The presence or absence of the circuit is of no consequence as long as the function is there.
3. An action which expresses a relationship between signal inputs and outputs. Some examples are AND, OR, AMPLIFIER, and INVERTER.
4. In positive logic, the most positive of two levels is a one, and the relative low is a zero. In negative logic, the converse is true.
5. a. Positive.   b. Negative.
6.

INPUTS		OUTPUT
A	B	C
−2 V	−2 V	−2 V
−2 V	+4 V	−2 V
+4 V	−2 V	−2 V
+4 V	+4 V	+4 V

Fig. A-10. Table of Combinations for Levels in Item 5.

7. a. The function is a positive AND.
   b.

Fig. A-11. Symbols for Positive AND.

8. a. The function is exclusive OR.
   b.

A	B	C
0	0	0
0	1	1
1	0	1
1	1	0

Fig. A-12. Truth Table for XOR.

9. $D = AB\bar{C}$

10.

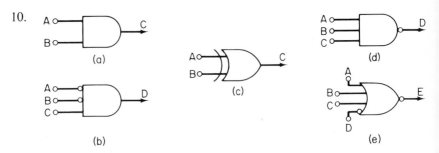

(a)

(b)

(c)

(d)

(e)

Fig. A-13. Logic Functions from Formulas.

11. a. $D = A + B + C$.   b. $\bar{C} = A + \bar{B}$.   c. $\bar{D} = \bar{A}B\bar{C}$.
   d. $C = \bar{A}B + A\bar{B}$.   e. $C = A + B$.   f. $D = ABC$.

12. a. The function is inversion (or negation).
   b.

Fig. A-14. Symbols for Inversion.

13. a. Negative logic.
   b. Positive logic.
   c. Logic negation.
   d. Logic negation from negative to positive combined with electric inversion.
   e. Logic negation from positive to negative without electric inversion.

14.

A		B	
L	1	1	H
H	0	0	L

Fig. A-15. Inverter with Logic Negation.

15.

A	B	C
0	0	1
0	1	1
1	0	1
1	1	0

Fig. A-16. NAND Symbols and Truth Table.

16. a.

Fig. A-17. Flip Flop.

b. Set, clear, and trigger inputs. One side and zero side outputs.

17. a. Apply a one to the set input.
    b. Apply a one to the clear input.
    c. Apply a one to the trigger input, or apply a one to both set and clear inputs at the same time.

18. a. Zero side = O V; one side = +6 V.
    b. Zero side = +6 V; one side = O V.

19. Temporary storage of a binary bit.

20. The flip flop latch has no trigger input, and it does not complement.

21. Both outputs will assume the same logic level for the duration of the inputs.

22.

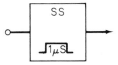

Fig. A-18. Single Shot with 1 μs Delay.

23. One μs. At the end of this period it automatically reverts to the quiescent state, which is equivalent to applying a clear pulse.

24. It provides a high (1) output as long as it is activated. It is activated when the input rises above the turn on potential, and remains activated until the input drops below the turn off potential.

25.

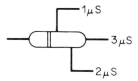

Fig. A-19. Time Delay with Three Outputs.

26. Figure 6-48 is a buildup symbol showing four identical flip flops. The clear and trigger inputs are common to all flip flops. Each flip flop has individual set inputs as well as individual outputs from each side.

27.

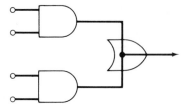

Fig. A-20. Two ANDS with Connector OR.

28.

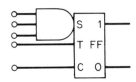

Fig. A-21. Combined AND and Flip Flop.

29.

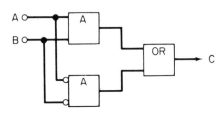

Fig. A-22. $C = AB + \bar{A}\bar{B}$.

30. a. $\bar{C} = (A + B)(\bar{A} + B)$.

31. a. Positive.   b. Negative.   c. Hybrid.
    d. Hybrid.

32. a. $C = AB$.   b. $\bar{C} = \bar{A} + \bar{B}$.   c. $C = \bar{A}B$.   d. $C = \bar{A} + \bar{B}$.

33.

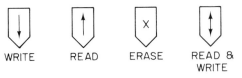

Fig. A-23. Four Types of Magnetic Heads.

Chapter 7

1. Any two opposite conditions that can be described by one and zero.

2. Any equation which expresses the relation between input and output signals.

3. AB, A·B, A × B, and (A)(B).

4. Vincula, parentheses, brackets, and braces.

5. AND, OR, and NEGATION.

6. a. Parallel.   b. Series.

7. By using the bubble state indicator on the input or output of a logic function symbol.

8. The AND becomes OR, and the OR becomes AND.

9.

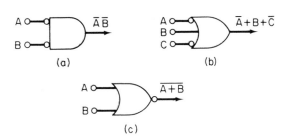

(a)   (b)

(c)

Fig. A-24. Logic Symbols from Boolean Expressions.

10. $\overline{AB}$ is the output of a NAND gate which has inversion on the output. $\overline{A}\,\overline{B}$ is the output of a two legged AND gate with inversion at both inputs.

11.

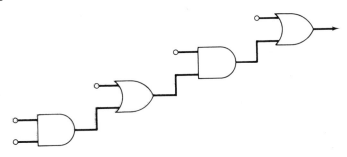

Fig. A-25. Fourth Order Logic.

12. a. Fourth.   b. First.   c. Second.   d. First.
    e. Third.

13. $1 = VW.\ 2 = VW + X.\ 3 = (VW + X)Y.\ 4 = (VW + X)Y + Z.$

14.

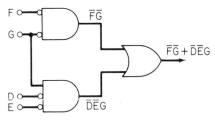

Fig. A-26. Logic for $\overline{FG} + \overline{DEG}$.

15. a. Sixth order.   b. $\{[(S + T)R + Q]P + O\}N$.

16.

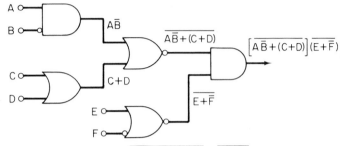

Fig. A-27. Logic for $[A\overline{B} + (C + D)](\overline{E + F})$.

17. $1 = AB\overline{C}$.

$2 = \overline{AB\overline{C} + D + E}$.

$3 = (\overline{AB\overline{C} + D + E})F$.

$4 = \overline{G}HI$.

5. $\overline{G}HI + J + K$.

$6 = (\overline{\overline{G}HI + J + K})L$.

$7 = [(\overline{AB\overline{C} + D + E})F] + [(\overline{\overline{G}HI + J + K})L] + \overline{M}$.

18. It eliminates unnecessary gates.

19. a. Absorption.   b. Idempotent.   c. Identity.

    d. Negation.   e. Identity.   f. Negation.

    g. Intersection.   h. DeMorgan's   i. DeMorgan's

    j. Distributive.   k. Associative.   l. Union.

    m. Complementary.   n. Complementary.   o. Absorption.

    p. Intersection.   q. Union.   r. Associative.

    s. Distributive.   t. Commutative.   u. Commutative.

20. $A(\overline{A} + B) = AB$ and $A + \overline{A}B = A + B$.

21. a. FG.   b. $KL + M$.

    c. D.   d. A.

    e. $ST + SV + RT + RV$.   f. $A + G$.

    g. FG.   h. B.

22. a. $(\overline{J\overline{K} + \overline{H\overline{J}}})L + M$.

    b. $L + M$.

    c.

Fig. A-28. Logic for $L + M$.

23. A minterm is the symbolic product of a given number of variables.
24. A minterm with one or more variables missing.
25. RTS, $\overline{R}\overline{T}\overline{S}$, $\overline{R}TS$, $R\overline{T}S$, and $RT\overline{S}$.
26. a, b, and d are correct minterms.
27. a, d, and e are correct minterm type terms.
28. An expression which is composed entirely of minterms and minterm type terms.
29. a. $\overline{T}\overline{V} + RS$.
    b. $A + \overline{B} + \overline{C} + \overline{D} + \overline{A}\overline{B}\overline{C}$.
    c. $AB + CD + EB + BC + \overline{D} + \overline{E}$.
    d. $ED + EF + RS$.
30. a. Eight.   b. 16.   c. 32.
31. One for each possible minterm.
32. a. 32.   b. Eight.   c. 16.
33.

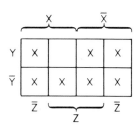

Fig. A-29. Plot for $\overline{Z} + \overline{X} + \overline{Y}$.

34.

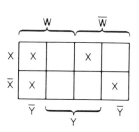

Fig. A-30. Plot for $W\overline{X}\overline{Y} + \overline{W}XY + \overline{Y}WX + \overline{X}\overline{Y}\overline{W}$.

35. a. $\overline{B}C + A\overline{B}\overline{C} + \overline{A}\overline{B}\overline{C}$.   b. $\overline{B}$.
36. a. $\overline{P}Q\overline{R} + \overline{Q}R + P\overline{Q}R + \overline{P}\overline{Q}R$.   b. $\overline{Q}R + \overline{P}Q\overline{R}$.
37. a. $JK\overline{L} + \overline{J}K\overline{L} + \overline{J}KL + \overline{J}\overline{K}\overline{L} + \overline{J}L + KL$.

b.

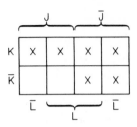

Fig. A-31. Plot for $JK\bar{L} + \bar{J}K\bar{L} + \bar{J}KL + \bar{J}\bar{K}\bar{L} + \bar{J}L + KL$.

c. $J + K$.

38.

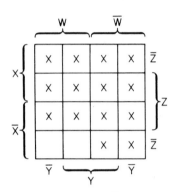

Fig. A-32. Plot for $\bar{W} + X + Z$.

39. $J + \bar{K} + L + \bar{M}$.

40. a. $\bar{H} + \bar{I}$.   b. $\bar{Q}S + \bar{P}S$.

41. a. $\overline{B + (A + \bar{C})(\bar{A} + C)} + \overline{A + C} + AB\bar{C}$.
 b. $\bar{A}\bar{B}C + A\bar{B}\bar{C} + \bar{A}\bar{C} + AB\bar{C}$.
 c.

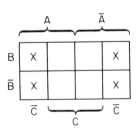

Fig. A-33. Plot for $\bar{A}\bar{B}C + A\bar{B}\bar{C} + \bar{A}\bar{C} + AB\bar{C}$

d. $\bar{A}\bar{B} + \bar{C}$.

e.

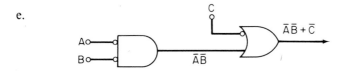

Fig. A-34. Equivalent of Fig. 7-69.

42.

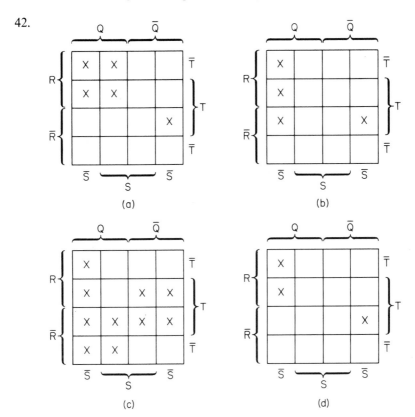

Fig. A-35. ANDed Diagrams.

The simplified expression is $QR\bar{S} + \bar{Q}\bar{R}ST$.

43. a. $J\bar{K}\bar{L}\bar{M}$.   b. $\bar{R}\bar{S}$.   c. $\bar{Q}R\bar{T} + \bar{Q}RS$.

Chapter 8

1. a. A group of computer circuits which perform a specific function.
   b. Encoder, detector, and adder.

2. A device for storage of a computer word of information.

3. A shift register is a storage register with shifting capabilities.

4. It must be cleared to all zeros.

5. A4 is the LSB and A1 is the MSB.

6.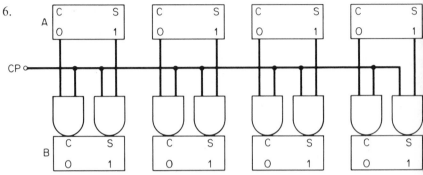

Fig. A-36. Simplified Forced Feed Transfer.

7. It eliminates the need for clearing the receiving register.

8. a. A serial input shift register.
   b. Because the data bits are arriving one at a time.

9. Because parallel transfers are faster than serial transfers.

10. Two registers are connected in series to accommodate the double length result.

11. a. It must be rounded off to a standard word length.
    b. The MSB of the extra register is sampled. If this bit is a one, one is added to the LSB of the accumulating register. If the bit is a zero, the contents of the extra register are discarded.

12.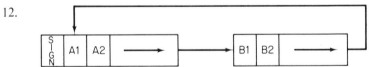

Fig. A-37. Double, Right, Arithmetic, End Around Shift.

13. Double, left, logical, end off.

14. By sensing the sign bit. If the sign bit is a zero, the number is positive.

15. It is spelled out in a instruction.

16.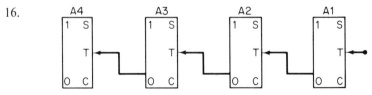

Fig. A-38. Serial Down Counter.

17. 15_{10}.

18. The pulses to be counted go to all stages, and the output of each flip flop goes to all higher order flip flops.

19. a. Maximum conditions $= 2^8 = 256_{10}$.
    b. Maximum count is $256 - 1 = 255$.

20. New count is 34_{10} after two complete cycles.

21. 512 μs.

22.

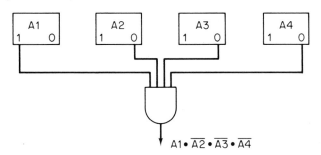

Fig. A-39. Detecting the Number Eight.

23. The AND gate is detecting the presence of the decimal number 12; the OR gate is detecting the absence of that number.

24. AND OUTPUT $= A1 \cdot A2 \cdot \overline{A3} \cdot \overline{A4}$.
    OR OUTPUT $= \overline{A1} + \overline{A2} + A3 + A4$.

25.

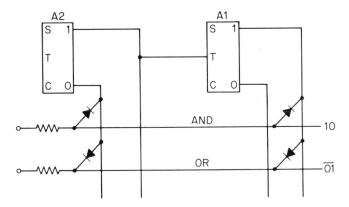

Fig. A-40. Diode Matrix Detecting 10 and $\overline{01}$.

26. 2048 or (8×2^8).

27. The treed matrix uses two logic levels of detectors.

28. 101.6 volts.

29. The coding tube, the comparator, and the binary coded disk.

30. a. One input = one, and the other = zero.
    b. Both inputs = one.

31. a.
    b.

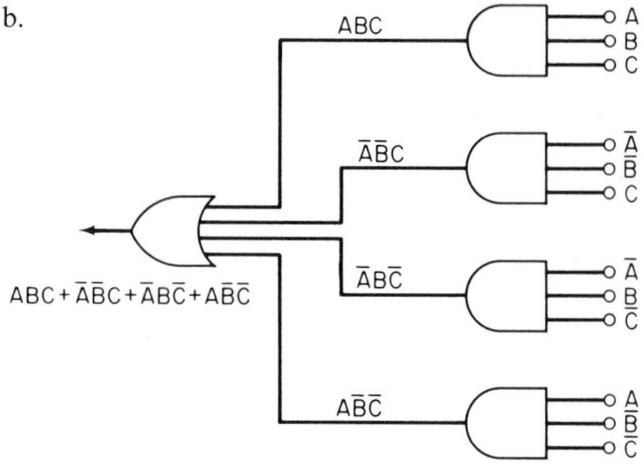

Fig. A-41. Summing Portion of a Full Adder.

32. Multiplication is a series of additions. Subtraction is addition of the minuend and the complement of the subtrahend. Division is a series of subtractions.

33. The adder produces a sum and a carry while the subtractor produces a difference and a borrow. The logic diagrams are essentially the same.

Chapter 9

1. Any device that can be used to store and retrieve information.

2. Mechanical, electronic, optical, and magnetic.

3. File and dynamic.

4. Punched cards and perforated paper tape.

5. a. Retrieve information from punched cards.
   b. Record information by punching holes in cards according to a fixed code.

6. An alphanumeric code for recording information on punched cards.

7. Four.

8. They fit the star-drive wheel which moves the tape.

9. Electronic.

10. A nonvolatile device will retain its information when power is removed.

11. a. The laser equipment is expensive.
    b. The stored information is inflexible.

12. Tapes, drums, disks, thin film, magnetic domains, and ferrite cores.

13. Non-return to zero (NRZ).

14. Serial access is too slow.

15. a. Channels can be accessed directly and at random.
    b. Information within a channel has serial access.

16. Tapes, drums, and disks.

17. Its relative high cost overrides its advantages.

18. The low reluctance makes them easy to magnetize. Their high retentivity enables them to retain information for an indefinite period of time.

19. Flux in one direction is a one; the opposite direction is a zero. When a one is read out, a zero is stored. Once magnetized the core can never again be neutral.

20. a. 32.   b. 64^2 or 4096.
    c. 12.   d. 32.
    e. 2048.   f. 32. One core on each plane.
    g. 32.   h. 32.
    i. 4096.

21. The zero is already there. The objective is to keep it there. The X line carries half current and the Y line carries half current. The inhibit line also carries half current, and it is in the opposite direction to the Y current. Inhibit current cancels the effect of the Y current, and the core remains in the zero state.

22. a. A pulse of current on the sense line.
    b. The absence of a pulse on the sense line.

23. The X and Y currents coincide in each core of the word. Each core receives full read current, which is sufficient to switch it from the one state to the zero state. The cores with ones are switched to zeros. The cores with zero remain in the zero state.

24. Storage cycle and retrieval cycle.

25. Read portion and write portion.

26. 0000 through 7777.

27. 64X and 64Y.

28. a. Words going into storage have a parity bit assigned. The parity bit will be such that all words being stored will have an odd number of one bits. As words are removed from storage, each word is checked for odd parity.
    b. Assigning and checking of parity occurs in the MIR.

29. 2, 3, R, D, H, Q, and U.

Chapter 10

1. The central memory, the arithmetic section, and most of the control section.

2.

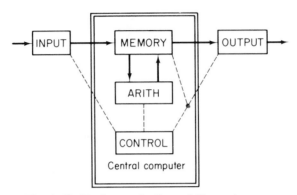

Fig. A-42. Equipment in Central Computer.

3. After both program and data have been stored in the central memory, the central computer can operate independently of the input and output sections.

4. It obtains instructions from memory, decodes them, obtains the data from memory, and guides the arithmetic section through each step of instruction execution.

5. He must set the starting address into the program address counter and push the start button.

6. After each instruction is obtained, the count in the program address counter is increased by one to provide the address of the next sequential instruction.

7. Program and data can be stored by operating toggle switches and push buttons. Information can be monitored by observing indicator lights which show the contents of key registers.

8. S, 3, 5, 9, 12, 14, and P.

9. In the MIR.

10. Odd.

11. Program cycle, operate cycle A, operate cycle B, and memory cycle.

12. a. Program cycle, operate cycle A, and two memory cycles.
    b. Program cycle, operate cycle A, and two memory cycles.
    c. Program cycle, operate cycle B, and two memory cycles.

13. Manual set, automatic count, programmed skip of one or two instructions, and a programmed JUMP to another section of the program.

14. Step 20 will be a jump instruction to divert control to a subroutine which calculates the square root table and stores it into central memory. The final instruction of the subroutine is another jump instruction to return control to step 21 of the main program.

15. Subroutine jump and return.

16. 1000_8.

17. One number is in the accumulator and the other is in the designated core location. The C(selected core) moves into MIR, where a parity check is performed. The A register is cleared, the data from MIR moves into the A register, and a copy is replaced in the same core location. The C(A register) are added to the C(accumulator) and the sum replaces the original accumulator contents.

18. 65665.

19. a. (MUL + DIV)TP4. b. (MUL + DIV)TP5.
    c. (CAD + ADD + SUB)TP6.

20. The accumulator cannot be cleared; therefore, CAD becomes identical to ADD.

21. SUB. The A register is not complemented and the SUB instruction becomes identical to ADD.

22. They require more time than is available in a normal cycle.

23. (MUL + DIV)MDTP4·STEP COUNT 0.

24. It eliminates end carry and speeds ths subtraction process.

25. At TP4 of either a MUL or DIV, command 5 complements the sign control flip flop if the A register is negative. At TP5 of a MUL or DIV, command 7 complements the sign control flip flop if the accumulator is negative. At TP8, command 15 complements the contents of the B register if the sign control flip flop is set.

26. At each MDTP2 when the signs of the A register and accumulator are the same.

27. 10 MHz.

28. 0.5 μs.

29. MUL·MDTP0·B6.

30. $\overline{B6}$.

31. The number of left shifts required to place a one in position A1.

32. Seven (one for each count plus one).

33. a. DIV·MDTP3·$\overline{\text{ACC SIGN}}$. b. DIV·MDTP3·ACC SIGN.

34. Overflow takes place when an arithmetic operation generates a word too large for the registers.

35. a and c cause overflow.

Chapter 11

1. Peripheral devices provide two way communications between the operator and the central computer.
2. a. Input.   b. Output.
3. Combined input/output represents the majority of the total system operating time.
4. Synchronous: entire operation is under computer supervision. All computation ceases until the I/O operation terminates.
   Asynchronous: Computer initiates the I/O operation, then continues processing data. The I/O operation is supervised by the I/O controller.
5. It interprets a control word which it receives from the central memory.
6. a. Three.
   b. Each I/O controller can operate only one peripheral device at a time.
7. Simple and complex. Simple devices operate in one direction only: either input or output. Complex devices perform both input and output operations.
8. a. Simple.   b. Complex.   c. Simple.   d. Complex.
   e. Simple.   f. Complex.
9. a. Output.   b. Both.   c. Output.   d. Both.
   e. Input.   f. Both.
10. The magnetic drum.
11. a. A card cycle is the distance (which represents a specific time) between the leading edges of the consecutive cards.
    b. A cycle point is the distance (or time) between two adjacent rows on a card.
12. a. It moves through the reader edgewise, face up, and 9 row first.
    b. It moves through the punch lengthwise, face up, and column 1 first.
13. Pins, brushes, and photocells.
14. The perforated paper tape drive provides inputs only. The magnetic tape drive provides both inputs and outputs.
15. Eight tape drives.
16. Use of the line printer.
17. It saves central computer time. The computer can write onto magnetic

tapes faster than the printer can print. After data is recorded on tape, tape and printer can be taken off line. The tape can then operate the printer and convert the data to hard copy without interfering with the central computer.

Chapter 12

1. Core to MIR, to A register, then through the adders.
2. Modifying data addresses.
3. a. Program counter.   b. Address register.
   c. I/O address counter.
4. In the process of branching.
5. a. S-3.   b. 4-9.   c. 10-15.   d. S-3.   e. 10-15.
   f. 4-9.   g. 10-15.
6. This machine has a right shift built into the add operation.
7. B register sign position.
8. 91, 41, SM, 101, and 77. 91 moves the C(Prog Ctr) into the memory address register; 41 clears MIR; SM starts memory; 101 clears the index interval and operation registers; and 77 clears the address register.
9. a. When the I/O interlock is set at PT10 of the HLT.
   b. An I/O is in progress.
10. 104 only.
11. The content of a specified index register is used to modify the data address portion of the instruction.
12. a. SUB.   b. 36.   c. 5.
13. a. 62.   b. 25_8.
14. Insert 0000 into bits S-3 of the instruction.
15. The machine has both positive and negative numbers. When subtracting numbers of unlike signs, the difference is larger than either number.
16. a. 12.   b. 6.   c. 16.5.   d. 51.5.
17. 15_{10}.
18. 80_{10}.
19. 1.010011100000110.
20. Addition is the only way to move information from the A register to the accumulator. Command 60 initiates this addition.
21. a. The C(Acc & B Reg) will be exchanged.
    b. 11.5.

22. C(Acc) is 1.101010101010101.
    C(B Reg) is 0.010101010101010.
23. a. 1.001010101010101.   b. 52.
24. a. 0.000101001110100.   b. 0.100101001001110.
25. a. Card reader.   b. 01.   c. Input.   d. 41_8.
26. The pause will last for about 198 μs. There are 33_{10} words to be transferred, and each word requires a 6 μs break.

Chapter 13

1. a. An instruction defines one step in a computer operation.
   b. A program defines all the steps in an entire procedure.
   c. Programming is the act of assembling instructions and data to cause the computer to perform a given task.
2. The three items for the machine are instruction location, the operation to be performed, and the location of the data to be used. The fourth item is the annotation, the programmer's note to himself as to why this step was used.
3. (1) X, (2) (X + H), (3) P(X + H), (4) P(X + H), (5) C, (6) C-P(X + H), (7) C-P(X + H), (10) C-P(X + H).
4.

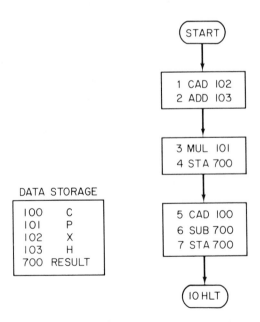

Fig. A-43. Coded Flowchart for In-Line Routine.

5. This provides for the automatic access operation of the program counter.

6. The unconditional branch and the decision making instructions such as test bits, compare, and branch on condition.

7. The address portion of an instruction contains the location of the necessary data.

8. The task can be accomplished with fewer instructions.

9. Problem analysis, organization, coding, and testing.

10. The problem analysis phase.

11. A mathematical statement which expresses the problem in its simplest form.

12. It does the job with a minimum of instructions and in minimum time.

13. Organization, coding, and testing.

14. The programmer must develop each path in order to provide for all possibilities.

15. The system flowchart shows all the major data processing steps. The program flowchart shows a flow of events guided by specific instructions.

16. Micro programming uses many instructions to define each step in relatively simple tasks. Macro programming uses a few instructions to call up large routines.

17. a.

18. Machine language is awkward, difficult, and prone to error.

19. a. Common business oriented language.
    b. It is a symbolic programming language.

20. A translator (assembly) program is loaded into a computer. The symbolic coded program is then loaded in as data. The computer translates the symbolic code into binary code, computes the address, and critiques the program.

21. Errors are always expensive, but the expenses are multiplied many times if the error is detected after trying a machine run. This type of error detection costs the time of both the programmer and the computer system.

index